Applied Climatology

STUDIES IN PHYSICAL GEOGRAPHY
Edited by K. J. Gregory

Also published in the series

Man and Environmental Processes
Edited by
K. J. GREGORY and D. E. WALLING

Geomorphological Processes
Edited by
E. DERBYSHIRE, K. J. GREGORY and J. R. HAILS

Ecology and Environmental Management
C. C. PARK

STUDIES IN
PHYSICAL GEOGRAPHY

Applied Climatology

A Study of Atmospheric Resources

JOHN E. HOBBS

*Senior Lecturer in Geography,
University of New England,
Armidale, New South Wales*

BUTTERWORTHS
London Boston
Durban Sydney Toronto Wellington

First published in Britain by Wm Dawson & Sons Ltd 1980
First Butterworths edition 1981

British Library Cataloguing in Publication Data

Hobbs, John Ernest
   Applied climatology.—(Studies in physical geography   ISSN 0142–6389).
   1. Climatology
   I. Title   II. Series
   QC981   551.6   80–41406

   ISBN 0–408–10737–5

Filmset in 10/12 point Times
Printed and bound in Great Britain
by Mackays of Chatham

Contents

For Helen, Matthew, Alison, and number three.

Acknowledgements

I am grateful to the following individuals and organizations for their permission to reproduce copyright materials:

Professor Reid A. Bryson, Institute for Environmental Studies, Madison, Wisconsin (Fig. 4.6); Dr William M. Gray, Colorado State University, Fort Collins (Fig. 4.1); Professor H. H. Lamb, Climatic Research Unit, University of East Anglia (Fig. 2.4); Dr Alison Macfarlane, National Perinatal Epidemiology Unit, Oxford (Fig. 3.11); Professor E. Palmén, University of Helsinki (Fig. 1.5); Dr R. A. Reck, General Motors Research Laboratories, Michigan (Fig. 2.10); Dr S. H. Schneider, National Center for Atmospheric Research, Boulder, Colorado (Fig. 1.3); Academic Press Inc., New York (Fig. 1.5); Air Pollution Control Association, Pittsburgh (Figs 6.5, 6.6); American Association for the Advancement of Science, Washington, DC (Fig. 2.10); American Meteorological Society, Boston (Figs 2.2, 4.8, 4.9, 4.10, 7.2, 7.3, 7.5, 7.6, 7.7, 8.1); Association of American Geographers, Washington, DC (Fig. 5.3); Australian Bureau of Meteorology, Department of Science and the Environment, Melbourne (Figs 3.3, 4.3, 7.4, 8.2, 8.3); Australian and New Zealand Association for the Advancement of Science, Sydney (Figs 2.3, 2.6); Clean Air Society of Australia and New Zealand, Sydney (Figs 2.8, 2.9); Elsevier Scientific Publishing Co., Amsterdam (Figs 2.5, 5.1, 5.4, 5.5, 5.6); Ferd. Dummlers Verlag, Bonn (Fig. 2.11, Table 2.2); IPC Magazines Ltd., London (Fig. 1.5); Macmillan Journals Ltd., London (Figs 3.11, 6.3); Controller of HMSO (Table 5.4); National Academy of Sciences, Washington, DC (Fig. 4.7); National Capital Development Commission, Canberra (Figs 3.1, 3.7, 3.8); Oriental Publishing Co., Honolulu (Fig. 1.6); Pergamon Press Ltd., Oxford (Figs 5.7, 6.1); Pitman Medical Publishing Co. Ltd., Tunbridge Wells (Fig. 3.9); Royal Meteorological Society, London (Figs 3.6, 4.5, 5.2, 5.8, 5.9, 5.10); Royal Swedish Academy of Sciences, Stockholm (Fig. 1.3); Swets and Zeitlinger B.V., Lisse (Figs 3.2, 3.5); University of Chicago Press, Chicago (Fig. 1.4); University of Toronto Press, Toronto (Fig. 6.2); World Meteorological Organization, Geneva (Figs 1.2, 2.7, 4.4, 6.4, 7.1).

It would not have been possible to complete the manuscript on time without the understanding and patience of my wife, Helen, and my young children, Matthew and Alison. Many people at the University of New England assisted in various ways with the preparation of this book. Thanks are due to Professors David Lea and Ian Douglas for making available the secretarial, cartographic, and technical facilities of the Geography Department. In particular, my thanks go to Carol, Tania, Bev, Jenny, and Cindy, all of whom struggled at various times with my handwritten drafts; to Rudi, Mike, and Neil for their excellent efforts with the figures, sometimes at very short notice; and to Pat for his assistance in sundry ways.

Armidale   Jack Hobbs
1979

Figures

Tables

Preface

Weather and climate are dynamic features of our environment and impinge upon all the acitivies of man to a greater or lesser degree. The ways in which the elements of weather and climate affect all forms of economic and social activity are now receiving increasing attention from climatologists. It is unfortunately true that the most familiar features of our surroundings tend to be those of which we take least notice; to which we give only passing thought; and which, because of their familiarity, tend to be ignored as items of interest. The atmosphere manifesting itself through weather and climate, provides frequent topics for superficial conversation, but rarely arouses a spirit of further enquiry.

Dyer (1975) has made the point that there are no more fundamental environmental components than the air we breathe, and the sunshine and rainfall that nourish our crops. Weather and climate are important factors in determining our day-to-day and longer term activities and life-styles. A generally increased awareness of our environment has led to development of the concept of a finite earth, and to concern with the problems posed by shrinking resources and increasing population. However, the concept of a finite atmosphere and limited atmospheric resources has caused little apparent concern and there is still a need for recognition and evaluation of the atmosphere as a resource. This raises many complex economic, social, and institutional problems and points to two basic ideas: first, that man is affected by the atmosphere and by information about the atmosphere; and secondly that man reacts to the atmosphere through his ability to make decisions (Maunder, 1971). To these can be added the role of man himself as a climatic factor of growing interest and concern, particularly with regard to his role as a cause of atmospheric pollution, perhaps leading to climatic change.

The need to consider the atmosphere as a resource to be managed requires some understanding of the fundamental nature of the atmosphere and weather processes, of the ways in which the atmosphere works, and why it works as it does. Therefore chapters 1 and 2 summarize the essential characteristics of the atmosphere, with particular emphasis on its variability in time and space, as a prelude to an investigation, in subsequent chapters, of the ways in which human behaviour and activities are related to weather and climate. The final part of the book (chapters 6, 7, and 8) then examines the various ways in which man is attempting to improve his use and knowledge of the atmospheric resource.

1
The Atmospheric System

1.1 THE SETTING (RESOURCE BASE)

1.1.1 Composition of the atmosphere

The atmosphere is a mixture of various gases, achieved by simple addition, without chemical reaction, held in an envelope around the earth by gravitational attraction. It contains amounts of water vapour, dust, and other liquid or solid particles which vary with time, location and altitude. Each of the gases, some permanent and some variable, can act quite independently of any others, and behaves in accordance with the gas laws.

Over 98 per cent of the total mass of air is nitrogen and oxygen, which are mixed in constant proportions up to an altitude of about 80 km. Nitrogen accounts for 75·5 per cent of a unit mass of air, oxygen for 23·1 per cent, argon 1·28 per cent and carbon dioxide about 0·045 per cent. There are also traces of other gases such as neon, helium, krypton, hydrogen and xenon. Air near the ground at the equator may also contain 2·6 per cent by volume of water vapour, but the colder air at latitude 70° might have as little as 0·2 per cent. At levels up to about 1000 km the proportion of oxygen increases. Gravitational sorting leads to larger proportions of the lighter gases such as helium and oxygen occurring at still higher altitudes.

There are about $2\cdot4 \times 10^{12}$ t of carbon dioxide in the atmosphere, but many times more in the oceans, which readily dissolve it, especially where the water is cold. The amounts of carbon dioxide actually present near ground level are affected by the temperature of the oceans, by the rate of plant photosynthesis, by man's use of fossil fuels, by respiration and by volcanic activity. Carbon dioxide is released into the atmosphere in large amounts through volcanic activity and man's burning of fossil fuels, and it is created by animal life. It is destroyed by photochemical processes in the upper atmosphere, is removed from the atmosphere in photosynthesis, and is dissolved in rain or sea water. The oceans act as a great reservoir of the gas, both storing and releasing it, depending upon water temperature.

Ozone is another important gas in the atmosphere and is a form of oxygen in which each molecule consists of three atoms instead of the usual two. Ozone occurs mainly in the region 15 to 60 km high, diffusing down from this region to the earth's surface. Concentrations at ground level vary with latitude and season but are always small. Even at about 20 km above the earth's surface concentrations are less than 6×10^{-4} g.m^{-3}, but this small amount is important because ozone absorbs the sterilizing ultraviolet rays from the sun's radiation which prevent photosynthesis, and which are dangerous to man.

The individual molecules of the various gases are continually cycling between different forms. Carbon dioxide is continually being converted into plant tissue in photosynthesis, later reentering the atmosphere through burning of wood or fossil fuels, or through plant and animal respiration; or it may be dissolved by the oceans and later released to the atmosphere. Oxygen goes through similar recycling and is also involved in photosynthesis and respiration. Likewise, water vapour is removed from the atmosphere as rain, runs in rivers to the oceans, and then evaporates back into the atmosphere. This water cycle may take only several days, but a complete cycle of a carbon dioxide molecule takes about 300 years, and for oxygen about 2000 years; while atmospheric nitrogen goes through the processes of fixation and liberation back into the atmosphere about every 10^7 years.

Current theories suggest that the earth was formed about $4 \cdot 5 \times 10^9$ years ago, with a primary atmosphere dominated by hydrogen. A secondary atmosphere, mostly of water vapour, methane, ammonia and hydrogen, began to form from gases emitted by molten materials of the Earth's interior. About 3×10^9 years ago cooling rocks gave off gases now emitted by volcanoes, namely water vapour, nitrogen and carbon dioxide. None of these gases contained free oxygen. It is still uncertain how oxygen came into the atmosphere, but there are several theories. A likely one involves the dissociation of water vapour by solar radiation. Further oxygen creation depended on photosynthesis, which gradually increased the oxygen content of the atmosphere until about one thousandth of the present oxygen concentration was reached at 10^9 years BP. At this stage the evolution of life went into higher gear, with the development of respiration. Much more complex plants became possible, permitting their extension into a wider range of environments. The increased vegetation then made further oxygen by photosynthesis. By about 4×10^8 years BP the atmosphere contained sufficient oxygen to allow life to exist on land. Then another rapid acceleration of the evolutionary process resulted in an increase of oxygen reaching present levels by about 3×10^8 years BP.

The history of our atmosphere is still a controversial subject, with many aspects disputed and requiring clarification. It is clear, however, than an important component of the atmosphere derives from plants, with the plant atmosphere relationship acting both ways. Oxygen concentration appears to be automatically held steady by the vegetation, while plants themselves evolve to suit the atmospheric conditions.

1.1.2 Structure of the atmosphere

The mass of air in the atmosphere is attracted by the earth's gravity, and so has considerable weight. At sea level atmospheric pressure is about 1013 millibars (101·3 kPa, or 1·013 kg.cm^{-2}). Air at any level is compressed by air above that level, so that the greater the altitude the less is the pressure. Air pressure is therefore about 700 mbar at 3 km, 500 mbar at 5·6 km and 300 mbar at 10 km above the earth's surface.

The density of air at mean sea level is about 1·2 kg.m^{-3}, but density depends on temperature, pressure and water content. The measures of sea level pressure and density suggest an atmospheric thickness of only about 8·4 km, but this ignores the fact that density is less at higher levels. The actual upper limit of the atmosphere is difficult to determine. The climatologist may be satisfied with a definition that includes those features relevant to climate at the earth's surface, but the physicist and chemist may be more concerned about the physical and chemical characteristics of the atmosphere and so accept a much higher limit. There is no atmosphere beyond 32 000 km, where the earth's gravitational attraction is exceeded by the centrifugal force caused by the earth's spin.

Not only does atmospheric pressure fall with increasing altitude, so does temperature (Fig. 1.1). Changes of temperature with height, or lapse rates, vary with time, place and

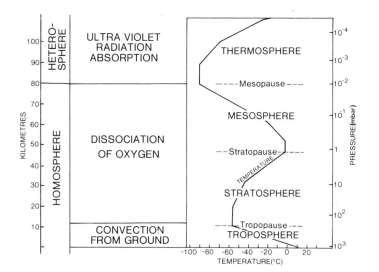

FIG. 1.1 TYPICAL FEATURES OF THE VERTICAL STRUCTURE OF THE ATMOSPHERE

elevation. In the lowest part of the atmosphere, winds and the buoyancy of air heated by the ground lead to continual stirring. Rising warm air is replaced by cooler air from above. Convection thus carries heat upwards, away from the ground. When a parcel of air rises it moves into regions of lower pressure and hence it expands and cools. Such cooling through expansion and the fact that elevated air is further away from its heat source at the ground, produce a temperature decrease, or positive lapse rate, with altitude in most situations (Fig. 1.2). Mean lapse rates in summer are about 5 °C.km^{-1} (5 mK.m^{-1}) in the lowest 2 km of the atmosphere, 6 mK.m^{-1} at 4 to 6 km and 7 mK.m^{-1} at 6 to 8 km. In winter and in inland areas lapse rates tend to be smaller. At low latitudes, because of stronger convection, lapse conditions usually extend to higher elevations, so that, paradoxically, the lowest temperatures in the atmosphere are over the equator.

A 'standard atmosphere' is a nominal relationship between altitude and temperature, representing a mean condition at a particular latitude. The most commonly encountered of several such relationships is that adopted by the International Civil Aviation Organization for use in calibrating aneroid barometers for measuring the altitude of aircraft. Its features are: an air temperature of 15 °C at sea level; a positive lapse rate of 6·5 mK.m^{-1} to 11 km; then isothermal conditions at −56·5 °C up to 20 km; and a negative lapse rate of

1·0 mK.m^{-1} to about 32 km. A negative lapse rate indicates increasing temperature with altitude, a situation known as a 'temperature inversion'.

The various features of atmospheric composition, heat convection, radiation absorption (section 1.2) and ionization tend to produce a stratified atmosphere. Ionization, the separation of electrical charges through bombardment of gas molecules by high energy

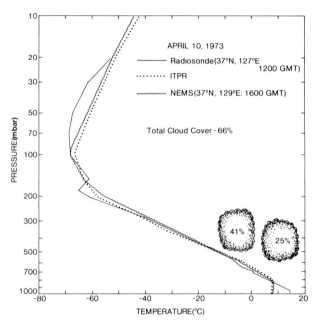

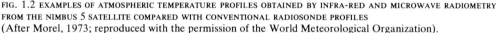

FIG. 1.2 EXAMPLES OF ATMOSPHERIC TEMPERATURE PROFILES OBTAINED BY INFRA-RED AND MICROWAVE RADIOMETRY FROM THE NIMBUS 5 SATELLITE COMPARED WITH CONVENTIONAL RADIOSONDE PROFILES
(After Morel, 1973; reproduced with the permission of the World Meteorological Organization).

cosmic radiation, occurs mainly at elevations of 80 to 700 km. It is negligible at heights where there are few molecules, and it cannot occur in the lowest levels of the atmosphere because the ionizing radiation is largely depleted by passage through the atmosphere. In the ionosphere electrical conductivity reflects radio waves back to the ground. It also leads to the optical effects of the aurora.

The atmospheric layer beyond 80 km is called the thermosphere because of the relatively high temperatures. These arise because the sparse gas is quickly warmed by the absorption of solar energy, but less readily loses heat. The mesopause separates the thermosphere from the mesophere beneath. The minimum temperature of about -90 °C at the mesopause results from the radiational loss of heat to space. The mesosphere beneath the mesopause is heated by the dissociation of oxygen.

At the stratopause dissociation heating is at a maximum, with temperatures reaching about 0 °C. The stratopause separates the positive lapse rate conditions of the mesosphere from the negative lapse rate conditions of the stratosphere.

The tropopause then separates the relatively sluggish conditions of the stratosphere from the much more turbulent troposphere, where lapse rates are again positive. The tropopause is found typically at altitudes of about 8 km over the poles, 18 km over the equator. The troposphere is characterized by a general decrease of temperature with height, increasing wind speeds with height, clouds, and vertical churning of the air by wind and convection. Most of the atmosphere's mass is within the troposphere, and most of the weather phenomena experienced by man at the earth's surface develop within this zone.

The lowest part of the troposphere can be further divided into three components: the 'friction layer' (or 'spiral layer' or 'planetary boundary layer'), the 'surface boundary layer' (or 'constant flux layer') and the 'laminar sublayer'. The friction layer consists of about the lowest 1 km of the atmosphere. Within it wind speed and direction are affected by the roughness of the ground. The surface boundary layer comprises the lowest few metres of the troposphere, where flows of heat, water vapour and energy may be regarded as vertical. The laminar sublayer is only millimetres thick and consists of air held almost stationary around all solid and liquid surfaces by molecular forces. It provides important thermal insulation.

The atmosphere is set in motion by input of radiation from the sun and the following sections consider what happens to this radiation and show how it generates atmospheric circulation.

1.2 ENERGY IN THE ATMOSPHERIC SYSTEM

1.2.1 Radiation

Energy from the sun is transferred to the earth by electromagnetic radiation. All objects emit such radiation. Most emit radiation of all wavelengths simultaneously, but particular wavelengths are usually dominant. The wavelengths of radiation from a body and the amount of energy transmitted depend upon the temperature of the body. Wien's Law states that the dominant wavelength of emission is inversely proportional to the absolute temperature (absolute zero or zero Kelvin equals -273 °C). The Stefan–Boltzmann Law states that the radiation of energy is proportional to the fourth power of the absolute temperature.

It follows that the sun, with surface temperatures at about 600 °K emits relatively short wavelength radiation, mainly within the range from 0·15 to 4 μm. Fifty-six per cent of the sun's radiation is within the range 0·38 to 0·77 μm. The human eye, which has evolved to benefit from the radiation chiefly available, is most sensitive to radiation in the band from 0·43 to 0·65 μm, from the longest wavelengths of blue light to the shortest wavelengths of red light.

The total amount of solar radiation intercepted by the earth and its atmosphere per day is roughly equivalent to the output of 10^8 large power stations, yet it is only one part in two billion of the total amount of radiation emitted by the sun. The amount received on a unit area facing the sun at the average sun–earth distance is called the solar constant, with a value of 1353 W.m^{-2}. The solar constant radiation is reduced by the obliqueness of the sun to the earth's surface, by cloud, by dust, and by the gases in the atmosphere. Therefore any area on the earth's surface actually receives radiation according to season, latitude, time of day and orientation of the surface, as well as being subject to variations in solar output and atmospheric absorption. These factors control the insolation, which is the solar radiation flux onto a unit area of ground surface beneath the atmosphere.

Solar radiation entering the atmosphere is partly absorbed, partly reflected, partly scattered and partly transmitted through to the surface (Fig. 1.3). Clouds greatly increase reflection and reduce transmission, but they usually absorb less than 3 per cent of the

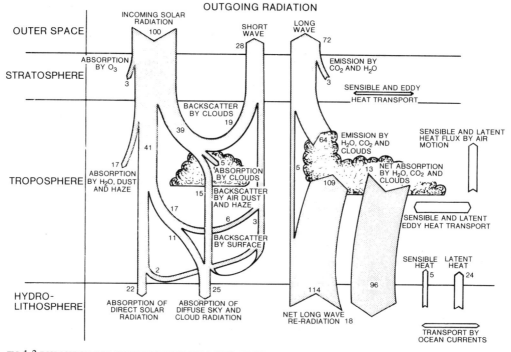

FIG 1.3 DISPOSITION OF RADIATION WITHIN THE EARTH-ATMOSPHERE SYSTEM
(After Schneider and Dennett, 1975; reproduced with the permission of the Royal Swedish Academy of Sciences and Dr S. Schneider).

incident solar radiation. Even in a cloudless sky only about 75 per cent of the radiation incident above the atmosphere actually reaches the ground. The depletion is caused by atmospheric gases and aerosols absorbing radiation and scattering some back to space. Absorption and scattering by the gases depend on the radiation wavelength and the kind of gas. Oxygen and ozone absorb the shortest wavelengths between about 0·02 and 0·29 μm, while shortwave radiation above 0·7 μm is mainly depleted by water vapour and carbon dioxide. Water vapour also absorbs in several narrow wavebands between about 0·9 and 2·1 μm. Gases tend to scatter shorter wavelengths than aerosols, because their molecules are much smaller. Shorter wavelength blue light is scattered more readily than red light, hence the blue of the sky.

Overall about 7 per cent of the incident solar radiation is scattered back to space from the atmosphere and about 18 per cent is scattered downwards to reach the earth's surface as diffuse radiation. The total solar radiation reaching a horizontal surface on the ground is called global radiation. It comprises the shortwave radiation direct from the sun, plus the diffuse radiation scattered by the atmosphere.

Seasonal variation of insolation depends on the changing earth–sun distance and on the tilt of the earth's axis. The slightly elliptical earth orbit means that the earth is closest to the sun in early January each year, and farthest away in early July. This might be expected to make the northern hemisphere winters warmer than those of the southern hemisphere and southern hemisphere summers warmer than those of the north. Such tendencies are masked, however, by the effects of the more extensive ocean in the south and by the fact that the atmosphere is in motion redistributing heat around the globe. In fact, the actual seasonal contrasts tend to be the reverse of what might be expected, with colder northern winters and warmer northern summers.

Smaller rates of average radiation flux onto areas at high latitudes result from two factors concerned with the sun being low in the sky: the small angle between the ground and the beam of sunlight means that radiation is shared over a larger area than would be the case with a vertical beam; and the obliquity of transmission through the atmosphere means that the solar radiation has to pass through a greater thickness of atmosphere to reach the surface, hence there tends to be greater atmospheric absorption.

The reflecting properties of clouds vary and some large and thick clouds may reflect as much as 90 per cent of the incident solar radiation, others as little as 30 per cent. A mean value for clouds is about 45 per cent. Part of the global shortwave radiation reaching the ground is also reflected upwards. The fraction returning in this way is called the surface albedo. The albedo of snow and ice is between 70 and 95 per cent. The albedo of oceans varies with cloudiness and with wave size from about 38 per cent when the sun is 10° above the horizon to about 4 per cent when the sun is overhead. Albedos of vegetated surfaces may range from 5 per cent for forests to 30 per cent for sparsely grassed areas. For soil the albedo depends on wetness, ranging from about 10 per cent for wet soil to 35 per cent for dry sand. The albedo of the whole earth–atmosphere combination, the planetary albedo, is usually estimated to be between 30 and 40 per cent, depending upon the extent of clouds, of polar ice and ocean surface, although recent views put the figure as low as 28 per cent.

Since the average surface temperature of the earth is about 285 K, it emits radiation in the infra-red spectral range from about 4 to 50 μm, with a peak near 10 μm. This flux of longwave radiation upwards from land and oceans is usually termed terrestrial radiation. Unlike solar radiation it is not reflected by atmospheric or terrestrial materials, but it can be absorbed, for example by clouds, and can create localized heating. Longwave radiation is also emitted from atmospheric gases and aerosols, both up and down. Temperatures in the atmosphere are usually lower than at the earth's surface, so downward atmospheric radiation is normally less than upward terrestrial radiation. The upward flux of atmospheric radiation to space helps to cool the earth, which would otherwise become steadily warmer as a result of solar radiation.

Most of the downward longwave radiation from the atmosphere comes from water vapour, which is mainly confined to the lower levels. Water vapour, clouds, and carbon dioxide also tend to absorb terrestrial radiation. Therefore, damp air near the ground effectively blankets the ground, keeping it warmer. A moist atmosphere warmed by absorption of terrestrial radiation, also radiates more effectively itself because of its higher temperature.

If all terrestrial radiation were absorbed by the atmosphere, the earth's surface would be far hotter than it is. Water vapour and carbon dioxide do not absorb radiation within the bands 3·2 to 4·3 μm and 8·5 to 11 μm. These radiation windows thus permit the escape of some radiation to space. It happens that the earth's temperature results in particularly strong radiation at such wavelengths, hence the windows permit an appreciable escape of heat. The overall input of radiation energy absorbed at the ground can be determined by subtracting combined upward fluxes (reflected global radiation, terrestrial radiation) from combined downward fluxes (global shortwave radiation, longwave atmospheric radiation). Clearly, this depends on all the factors which influence the constituent radiation fluxes: elevation of the sun, cloudiness, turbidity, albedo, the temperature and dryness of the atmosphere, and the altitude. Net radiation energy input from the sun is available at the ground for important processes such as photosynthesis, evaporation of water, and heating of the earth and air.

Global shortwave radiation dominates in the daytime, so the net radiation flux is then downwards. At night upward longwave fluxes are dominant. Lower nocturnal temperatures mean that terrestrial radiation is less than in the daytime, but it still remains greater than the counter atmospheric radiation. Therefore, at night the net radiation flux is upwards, cooling the ground. Cloud exerts a major influence on the rate of nocturnal cooling. Measurements in Sydney, New South Wales, in April, for example, have shown a fall of 2·3 °C between 6 and 9 p.m. with an overcast sky, but of 6·2 °C under clear conditions.

The net radiation of the earth–atmosphere system as a whole must be zero. A positive flow of net radiation would result in steadily increasing earth temperatures, and vice versa for a negative flow. The overall equality of incoming and outgoing radiations conceals variations over the globe. Net radiation values are particularly high at the surface of an ocean, because of low albedo and the reduced terrestrial radiation from a surface cooled by evaporation, and there is a net income in low latitudes but a net loss in high latitudes.

Higher temperatures at the equator mean that the outward terrestrial radiation at ground level is about twice that at the poles; but the input of solar radiation is also considerably greater in low latitudes. The annual shortwave input of radiant energy exceeds the longwave loss at latitudes less than about 40°, with a corresponding deficit at higher latitudes. This net heating difference in turn accounts for the temperature differences between latitudes. These then cause the poleward movements of heat, and hence the characteristic features of atmospheric and oceanic circulation develop, to produce in turn the variable features of weather and climate.

1.2.2 Energy balances

The atmosphere itself is a net loser of radiation at all latitudes, so there must exist a two-way heat transfer: from the earth's surface to the atmosphere, and from the equator to the poles. Atmospheric circulation and weather are created in response to these energy imbalances. Energy is carried polewards as sensible heat in warm winds and warm ocean currents, and as latent heat in wet winds; heat absorbed during the evaporation process is transported by the winds and later released during condensation (Fig 1.4).

About 60 per cent of the total heat transfer across circles of latitude is in the form of sensible heat. Oceans may account for at least 25 per cent of the total transfer. Latent heat flux accounts for only about 25 per cent of the total meridional (across latitude circles) energy transfer in the southern hemisphere and less in the northern hemisphere. In the zone between 10° N and 10° S latent and sensible heat transfers tend to be in opposite directions. Much of the latent heat released by water vapour carried into the zone by low

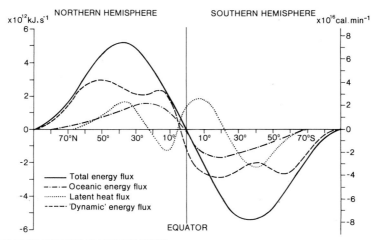

FIG. 1.4 ANNUAL MEAN NORTHWARD ENERGY FLUXES
(After Sellers, 1965; reproduced with the permission of the Univeristy of Chicago Press).

level air currents is transferred to higher levels in the atmosphere as sensible heat, which is then transported from the equatorial zone.

In Europe most of the available radiative energy is used for evaporation. In Asia and Australia, particularly the latter, most of the energy goes into warming the air. In the case of the Indian Ocean and indeed for other major oceans, about 90 per cent of the available energy is used for evaporation. The advective component is very small in relation to the total energy balance of the ocean. In Antarctica and the Arctic Ocean the mean sensible heat flux is from the air to the ground and there is a negative radiation balance.

1.3 ATMOSPHERIC CIRCULATION

1.3.1 Balance in the system

It is useful to remember that the whole atmospheric system has to operate within certain physical constraints. The need for redistribution of heat energy has already been mentioned. In a similar way moisture has to be redistributed by the circulation of the atmosphere, so that patterns of precipitation and evaporation are maintained. The average atmospheric water content is equivalent to a global rainfall of only 25 mm, so redistribution is clearly necessary to produce the observed rainfall totals. The average annual rainfall over the earth is about 1000 mm, hence there are about forty evaporation-precipitation cycles each year. Further, since there are no long-term increases or decreases of atmospheric pressure at any locality, this implies that the mass of air at any point does not change significantly when averaged over a long period, and for every transfer of air there must be a return.

In addition the atmospheric system is subject to numerous other influences, including the effects of friction between the earth and the atmosphere; varying radiative properties of the earth's surface; varying thermal effects because of the different physical and chemical properties of land and water; mechanical effects such as disturbance of airflow by relief features; and the effects due to the earth's rotation. Radiative, thermal and mechanical parameters are all subject to latitudinal and longitudinal variations in the irregular nature of the earth's surface. In this respect it is important to recognize the differences between the northern and southern hemispheres, with the ocean domination of the latter having a major influence on heat storage, albedo and friction effects.

1.3.2 Circulation features

The largest scale of atmospheric motion is the three-dimensional general circulation. This is the totality of flows of air, heat, and moisture around the globe. Global winds are made up of fluctuating local winds at all scales of motion, near the ground and high in the troposphere, varying with time and location. This complexity may be clarified by considering average conditions idealized into simple patterns. Many models have been produced to portray the basic features of the general circulation of the atmosphere based on accumulated observations and on physical and mathematical modelling.

It is conventional to discuss the basic features of the general circulation by reference to mean conditions in January and July (Figs 1.5 and 1.6). The patterns of wind and pressure are roughly the same in both northern and southern hemispheres, although differences do occur because of the different distributions of land and sea. Three main interconnected units are usually identified for each hemisphere, with strengths varying from place to place and season to season. The three units are the Hadley Cell in low latitudes, the Ferrel westerlies in mid-latitudes, and a much less organized or defined unit or zone of mixing in high latitudes (Fig. 1.5).

The intertropical convergence zone (ITCZ) is located at the confluence of the Hadley Cells in the two hemispheres. The zone is not uniform in width, depth or latitudinal position. In July it is centred north of the equator, generally farther away from the equator than its corresponding position in the southern hemisphere in January. In fact the pattern of ocean surface temperatures, which seems to have an important influence on the locations of the ITCZ, is such that in January it remains centred just north of the equator over all except the Indian Ocean and the westernmost portions of the Atlantic and Pacific Oceans. The ITCZ over the oceans is usually marked by the occurrence of lines and clusters of towering cumulonimbus clouds, where the latent heat transported by the surface trade winds is converted to sensible heat.

The descending branches of the Hadley Cells are associated with subtropical areas of high pressure and stability. This stability limits cloud growth and helps to maintain the vast

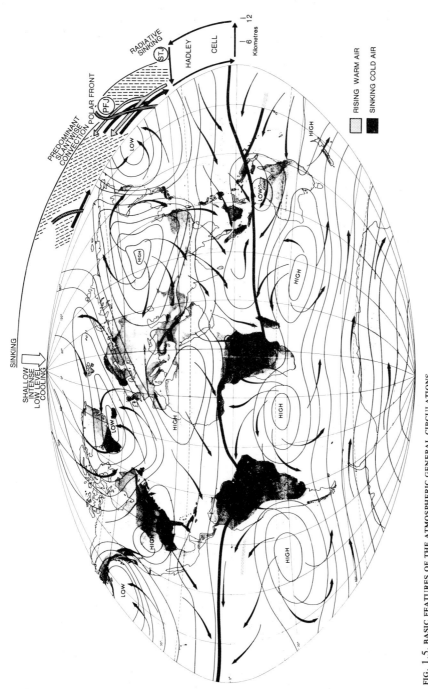

FIG. 1.5. BASIC FEATURES OF THE ATMOSPHERIC GENERAL CIRCULATIONS
(After Palmén and Newton, 1969, and Chandler and Musk, 1976; reproduced with the permission of Academic Press Inc., Professor E. Palmén, and I.P.C. Magazines Ltd.) The vertical section shows the situation for the northern winter. In summer major features are displaced polewards, so that the area affected by the polar anticyclone, polar easterlies and polar air currents contracts, while the area affected by tropical air expands.

subtropical deserts. Major areas of high pressure tend to be centred over relatively cold regions, such as oceans in summer and continents in winter. The high pressure areas in the northern hemisphere demonstrate this seasonal pattern quite well, but in the southern hemisphere are located over the oceans all year round with winter extensions across the southern continents. Variations in position, strength and shape of the main centres of high pressure in the subtropical zone are important in the general circulation of the atmosphere,

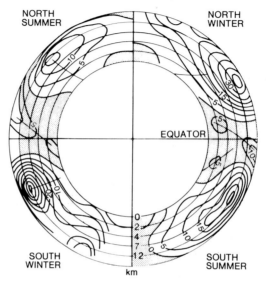

FIG. 1.6. MEAN ZONAL WINDS IN THE TROPOSPHERE
(From Chang, 1972, after Mintz, 1954; reproduced with the permission of the Oriental Publishing Co.) Isotachs in units of m.s^{-1}. Shaded areas show easterly flows.

since the air moving away from the centres comprises a large part of the entire surface air circulation.

The disposition of high pressure centres in the subtropics of both hemispheres and the tendency for winds to spin clockwise out of northern hemisphere highs and anticlockwise out of southern hemisphere highs, results in mainly easterly winds in low latitudes between the two high pressure zones. The moisture-laden trade winds thus emerge from the oceanic high pressure cells and converge on the ITCZ from north-east in the northern hemisphere and from south-east in the southern hemisphere. The trade winds are particularly evident on the eastern sides of the oceans and tend to be stronger in the southern hemisphere.

Westerly winds dominate the atmospheric circulation polewards of the subtropical high pressure zones. In this zone there are frequent occurrences of migrating high and low pressure centres so that circulations are much more complex than those characteristic of low latitudes. In the middle and upper troposphere of middle latitudes the flow pattern tends to be in a series of waves rather than cellular. The strength of the westerlies increases towards the tropopause where concentrations of maximum winds may form jet streams with speeds up to 500 km.h^{-1}. There are commonly two jets in each hemisphere: the subtropical jet on the upper poleward edge of the Hadley Cell, and another related to the discontinuous polar front of middle latitudes, where the major mid-latitude and polar circulation units interact. Interest in such upper air flows comes from their influence on surface weather, and from their relative simplicity and regularity for analysis and forecasting purposes.

In high latitudes the troposphere is only about half as deep as in the tropics and circulation systems tend to be shallow and short-lived. The picture of atmospheric circulation in polar regions is still confused. Data are generally inadequate, and in Antarctica the overall nature of the circulation is easily obscured by temperature inversions and by very strong local winds. Radiative cooling and sinking seem to dominate over the poles, with subsiding air being replaced at high levels by air rising at the polar front.

All schemes of the general circulation are necessarily highly simplified, but they do convey impressions of the broad climatic patterns of the world. General circulation models fail to indicate the unsteadiness of the atmosphere and its characteristic seasonal fluctuations. There are wide variations in the locations and intensities of all features of the atmospheric system, right down to the most localized scale, at which level uncertainty is most pronounced.

Theoretical studies of the atmospheric system, using laboratory equipment and high-speed electronic computers, have been able to simulate many features of the observed circulation, but there is still a long way to go before the complex mechanisms of the general circulation are fully understood. Lorenz (1970) has emphasized that time-averaging over a year, the basis of most approaches to the problem, tends to obscure many of the more interesting features of the system. In particular, it obscures the significance of the vast monsoon circulations of south and south-east Asia, and it virtually ignores the intricacies of the smaller scale subsystems which actually provide the weather experienced at any locality. Therefore, in the next chapter consideration will be given to some of the features of atmospheric variability in time and space.

SUGGESTIONS FOR FURTHER READING

ATKINSON, B. W., 1978, 'Topics in dynamical meteorology: 1: weather, meteorology, physics, mathematics.' *Weather*, 33, pp. 2–8.

FLOHN, H., 1970, 'Climatology—descriptive or physical science?' *WMO Bull.*, 19, pp. 223–9.

HAMILTON, M. B., 1976, 'The south Asian summer monsoon.' *Progr. in Geogr.*, 9, pp. 147–203.

JOHNSON, D. H., 1970, 'The role of the tropics in the global circulation.' In *The Global Circulation of the Atmosphere*, (Royal Meteorological Society, London), pp. 113–36.

KONDRATYEV, K. YA., 1971, 'Interaction between dynamic and radiative processes in the global circulation of the atmosphere.' *WMO Bull.*, 20, pp. 78–85.

KUTZBACH, J. E., 1976, 'The nature of climate and climatic variations.' *Quatenary Research*, 6, pp. 471–80.

PANOFSKY, H. A., 1978, 'Topics in dynamical meteorology: 2: hydrodynamics (1).' *Weather*, 33, pp. 42–8.

—1978, 'Topics in dynamical meteorology: 2: hydrodynamics (2).' *Weather*, 33, pp. 80–6.

—1978, 'Topics in dynamical meteorology: 3: thermodynamics (1).' *Weather*, 33, pp. 121–6.

—1978, 'Topics in dynamical meteorology: 3: thermodynamics (2).' *Weather*, 33, pp. 172–8.

PEARCE, R. P., 1975, 'Tropical meteorology—its economic importance and scientific challenge.' *WMO Bull.*, 24, 147–56.

WALKER, J. M., 1972, 'Monsoons and the global circulation.' *Weather*, 27, pp. 349–55.

2
The Variable Atmosphere

It is possible to identify atmospheric circulation systems at all time and space scales. In all cases there are complex interrelationships between mass and energy flows, pressure distributions and resultant weather features. In view of the continuity of changes of pressure and wind, any separation of the orders of atmospheric circulation is somewhat arbitrary. Although such separation is convenient and widely used in climatological work, we should never lose sight of the essential unity of the atmospheric system.

Variability in time and space is an inherent characteristic of the atmosphere and hence of climate. Maunder (1978a) has emphasized that the atmosphere is a variable resource, and that this variability can be evaluated at different scales which are useful to planning and management.

2.1 SCALE VARIABILITY

Averaging, when used to identify the overall patterns of the general circulation of the atmosphere, conceals the numerous relatively mobile features of smaller scales. A major division can be made into the features of pressure and wind making up the secondary circulations, which in turn contain the smaller-scale features of the tertiary circulations.

Secondary circulations are components of the general circulation induced by horizontal temperature differences within the troposphere and/or dynamically by convergence or divergence of air flows. Migratory high and low pressure systems of middle latitudes, such as might be identified on the daily weather maps seen on television or in newspapers, are characteristic secondary circulation features. Secondary circulations are changing all the time. Each low or high pressure system forms, matures and disappears to be replaced by others and so adds to the local variabilities of our weather.

Tertiary circulations form part of the chain of decreasing scale and increasing irregularity of motion progressively degrading available solar energy. They are driven by winds in the secondary circulations, by local differences of topography and by local temperature contrasts, such as that between land and sea. Local winds in turn generate large eddies and turbulence around surface irregularities.

Meteorologists commonly recognize the scales of atmospheric motion systems shown in Figure 2.1 The categories identified are usually used in reference to typical weather systems, but they may also be applied to other organized weather phenomena such as areas of precipitation or cloud cover. Identification of phenomena at the different scales depends

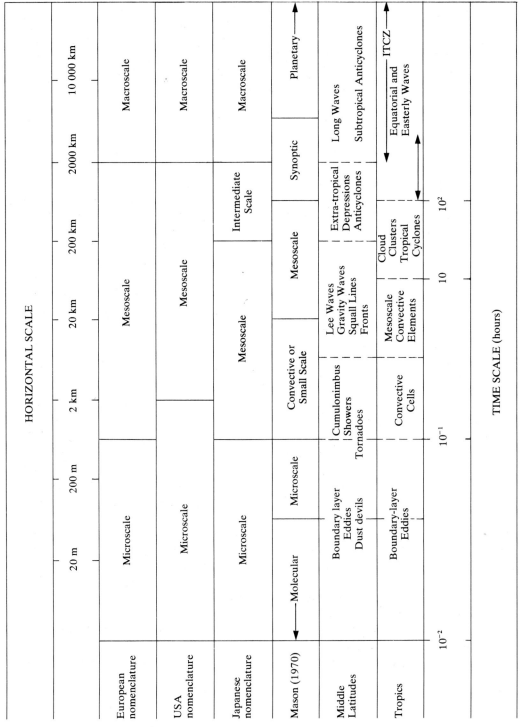

FIG. 2.1 SCALES OF ATMOSPHERIC MOTION SYSTEMS
(After Mason, 1970; Orlanski, 1975).

on the density of observational networks and on the particular analytical methods used. In practise a more or less complete spectrum of weather systems exists.

Meteorologists have tended to study the atmosphere from two points of view characterized by greatly differing space and time scales. Micrometeorology deals with atmospheric dynamics having space scales of several metres and time scales of the order of a minute, and is therefore concerned with the immediate environment in which man lives. At the macro-scale meteorologists have been concerned with the large-scale dynamics of the atmosphere with space scales greater than 500 km and time scales of the order of a week to ten days (Orlanski, 1975).

Macroscale and microscale meteorology together encompass many atmospheric processes, but there is still a large number of important phenomena occurring in the atmosphere with spatial and temporal scales intermediate between these two categories. The term mesoscale, as it is used today, defines all the intermediate states between the macroscale and microscale. There are no precise limits to the size of features which can be described as mesoscale, various authors specifying ranges from 1·5 km to about 800 km. The usually accepted size range is 15–150 km (Atkinson, 1970). Atkinson proposed a three-fold grouping of mesosystems into: topographically induced mesosystems; free-atmosphere convective and associated mesosystems; and other free-atmosphere mesosystems. Topographically induced mesosystems include, for example, lee waves and lee depressions, and sea breezes and lake breezes. Typical free-atmosphere convective mesosystems are the cells which form in cumulus and stratocumulus clouds over vast areas of ocean in response to heating from below. Mesoscale patterns are also apparent in non-convective rainfall associated with mid-latitude fronts and with tropical rain storms, suggesting the relatively common occurrence of mesoscale motion systems.

Similar scale categories have been suggested for climatic systems, as opposed to those of atmospheric circulation. The characteristics upon which the spatial systems of climate may be based are not as clear as in the case of motion systems. At the largest scale of climatic system the global wind belts are generally identified with reference to mean monthly streamlines, resultant winds, and measures of wind constancy. Regional macroclimates are usually delineated in terms of static distribution of climatic elements such as temperature and precipitation parameters. Local climates and microclimates are harder to distinguish.

According to Barry (1970) a distinctive local climate is the result of pronounced geographical features, such as relief, slope, aspect and surface type, which combine to modify large-scale energy and moisture budgets and air entering the area from elsewhere in mesoscale motion systems. Clearly, there is an endless range of possibilities. In defining microclimate horizontal scale is rarely mentioned and suggested vertical scales vary considerably from the lowest 2 m, to a level at which the effects of the immediate character of the underlying surface cannot be distinguished from the general local climate (Huschke, 1959). The Meteorological Glossary (Meteorological Office, 1972) provides a reasonable guide with a definition of microclimate as 'The physical state of the atmosphere close to a very small area of the earth's surface, often in relation to living matter such as crops or insects'.

2.2 CLIMATIC CHANGE

The world's climates have changed in the past, are changing now, and there is every reason to expect that they will change in the future. Recent concern with the impact of human activities on the environment has included an upsurge of interest in climatic change at all scales, from local to global. Numerous discussions have been published in the general press and in professional journals, with approaches ranging from completely theoretical to completely descriptive, and from detailed analysis of certain aspects of the problem to general speculation (Bryson, 1974a). Much discussion, even controversy, has been generated by various hypotheses and statements as to the likely onset of rapid climatic change (e.g. Bryson, 1973; Alexander, 1974; Lamb, 1974; Ponte, 1976). The importance of climatic changes to the world's future and of man's dependence on a stable climate have been highlighted in the proliferation of recent books about climatic change (e.g. Winkless and Browning, 1975; Gribbin, 1976a, 1978; Ponte, 1976; Bryson and Murray, 1977).

Bryson (1975) has summarized the lessons of climatic history as:

1. Climate is not fixed.
2. Climate tends to change rapidly rather than gradually.
3. Cultural changes usually accompany climatic changes.
4. What we think of as normal climate, at present, is not normal in the longer perspective of centuries.
5. When the high latitudes cool, the monsoons tend to fail.
6. Cool periods of earth history are periods of greater than normal climatic instability.

The major causes of climatic variability can be conveniently grouped into two basic categories. The first category includes internal causes which involve exchanges of energy between the atmosphere, hydrosphere, lithosphere, and cryosphere. These causes are self-stimulating mechanisms in the climatic system, influenced by the existing climatic state. The second category involves causes which are external to the system, since they are not considered to be influenced by the climatic state (Bach, 1976). External causes include variations in the orbital characteristics of the earth, fluctuations in solar emission, changes in atmospheric aerosol and carbon dioxide contents, and changes in the land surface and the heat budget (Kellogg and Schneider, 1974).

We still lack an entirely satisfactory theory for climatic change, although it is known that factors such as those mentioned above must be involved. It is ultimately changes in the heat balance of the earth–atmosphere system that cause climate to change and it is possible that man-produced pollutants could bring about such changes. Unfortunately, analysis of the relative effects of the various factors is complicated by the complex feedback mechanisms between various geophysical processes that suppress or amplify the impact of heat balance changes on climate (Bach, 1976, see Table 2.1). Assessment of the relative effects of the internal and external causes of climatic change and of the respective roles of natural and anthropogenic influences remains the important problem (Schneider and Dennett, 1975).

The paleoclimatic history of the earth is summarized in Table 2.2. Primary cycles can be distinguished, with the development of major ice ages about every 100 000 years and a

Table 2.1
THE ENERGETICS OF LARGE-SCALE CLIMATIC CHANGES (after Bach, 1976)

		Global, TW		W.m^{-2}	
External Parameters	Solar constant	173 000		340	
	Input of earth and atmosphere	123 000		241	
	Net radiation, earth surface	52 000		102	
	Geothermal heat	32		0·063	
	Volcanic dust, stratosphere (radiation deficit)	100–300			
	Antarctic ice surges (including melting)	50–100 .10^{6}km^{-2}			
Internal Parameters (with non-linear feedback)	Absorption in the atmosphere (short wave)	45 000		88	
	Production of available potential energy	1 200		2·4	
	Change of cloudiness (1%)	350		0·67	
	Change of evaporation equatorial oceans	300		0·59	
	Photosynthetic processes	92		0·18	
	Change of snow cover	110		0·22	
	Change of Arctic Sea ice area	50 .10^{6}km^{-2}			
		1970	*2000*	*1970*(mW.m^{-2})	*2000*
Anthropogenic parameters	Increase of carbon dioxide	+1·5	+2·4	3	5
	Energy production	+8	+40	15	78
	Savanna bushfires (direct heat input)	+3		6	
	Tropospheric dust from industry, cities	+1·7	+2·5	3	5
	Tropospheric dust from vegetation destruction	+5	+6	10	12
	Water consumption (evaporated)	+140	+390	270	765
	Conversion of tropical rainforest into cropland (changing heat budget)			−17 .10^{6}km^{-2}	

termination period of 10 000 years. Secondary oscillations are apparent with glacial growth over 20 000–30 000 years and retreats in 1000 years. Smaller time-scale, but significant, fluctuations have also been identified, such as the Little Ice Age from 1500–1700 AD, the northern hemispheric warming trend from the 1880s to the 1940s, and the southern hemisphere warming trend from the early 1950s. Such small-scale fluctuations do not occur uniformly over the globe, they tend to be more pronounced at higher latitudes, and temperature trends may even be in opposite directions at sites relatively close together (e.g. St Louis and Edmonton, Fig. 2.2).

Examination of recent climatic history gives what can best be described as a confusing picture. Global average temperatures rose by about 0·5 °C from the early 1880s to the early 1940s, and have since fallen by 0·2 to 0·3 °C. Such temperature fluctuations are assumed to reflect a systematic change of the planetary heat budget, and have apparently been accompanied by changes of the large-scale atmospheric circulation and other climatic elements. It is likely that the bulk of the fluctuations can be ascribed to natural causes, particularly the varying load of volcanic dust. Part, however, may be related to human activities, especially to an increase of atmospheric pollutants. Many authors have attributed the warming period to increasing atmospheric carbon dioxide concentrations, but the subsequent cooling requires a new hypothesis, perhaps the effects of increased particulate matter (e.g. Mitchell, 1972; Kukla and Kukla, 1974; Lamb, 1974; Barrie et al., 1976). These theories are still difficult to prove or disprove, because the observed fluctuations are not outside the range of possible natural variability. In addition, global average figures

Table 2.2
PALAEOCLIMATIC HISTORY OF THE EARTH
(after Bach, 1976; reproduced with the permission of Ferd. Dummlers Verlag)

Pre-Cambrian Era	(5000–550 MYBP[a])
	Most widespread glaciation ca. 700 MYBP
Paleozoic Era	(500–200 MYBP)
	During the late Paleozoic (300–250 MYBP) Gondwana was glaciated for 30–50 MY. During most of this era the earth was, however, warmer than today because 80–90% of the time the poles were ice-free.
Mesozoic Era	(200–60 MYBP)
	Much warmer than today. Annual mean surface temperatures were 8–10 °C near both poles and 25–30 °C in the tropics.
Cenozoic Era	(60 MYBP–Present)
Tertiary Period	(60–1 MYBP)
	Main feature of this era was the gradual decrease of mid-latitude temperature from about 20 °C above the present temperature in the Eocene to about 10 °C in the Pliocene. Antarctic glaciation began during the Pliocene (5 MYBP). Mountain glaciers in the Sierra Nevada, Iceland, and Greenland were formed about 2·5 MYBP.
Quaternary Period	(1 MYBP–Present)
	The Pleistocene was characterized by a sequence of glacial and interglacial periods. We can distinguish:
	Primary Cycles consisting of 4–5 major glaciations
	development in 90 000–100 000 years
	termination in 10 000 years
	Secondary oscillations with
	glacial growth in 20 000–30 000 years
	retreat in 1 000 years
	The last glacial maximum in Europe, the Würm, and in North America, the Wisconsin, was 20 000–18 000 YBP.
	The present interglacial (Holocene) began 12 000–10 000 YBP and reached a climatic optimum about 5 000 YBP.
	Tertiary fluctuations
	Mild climate in early Middle Ages, 800–1 000 AD
	Iceland and Greenland cultivated by Vikings
	Little Ice Age in 1 500–1 700 AD
	Worldwide warming of about 0·6 °C
	from 1880s to 1940s
	Northern hemispheric cooling by about 0.3 °C
	since the 1940s with many trend reversals
	Southern hemispheric warming by about 0·5 °C
	since the 1950s at Australian, New Zealand and Antarctic stations

[a] MYBP = million years before present

tend to mask considerable regional departures from the overall trend, and there are some indications that since 1965 global average temperatures have stopped falling, and may even have increased slightly since about 1970 (Angell and Korshover, 1977).

Broecker (1975) reported that the $^{18}0$ record in the Greenland ice suggests that the present northern hemisphere cooling is one of a long series of similar natural climatic fluctuations, which has more than compensated for any warming effect produced by carbon dioxide. By analogy with similar cooling trends in the past, the present trend should bottom out over the next decade or so. Then the carbon dioxide effect might become a

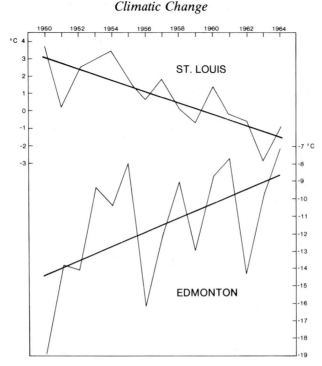

FIG. 2.2 TIME SECTIONS OF WINTER MEAN TEMPERATURE AT ST LOUIS AND EDMONTON, WITH TREND LINES FOR THE PERIOD 1950 TO 1964
(After Van Loon and Williams, 1976; reproduced from the *Monthly Weather Review*, p. 366, with the permission of the American Meteorological Society).

significant factor, so that by the next century we may experience global temperatures warmer than those of the last 1000 years.

Australian and New Zealand temperature data generally do not reflect the trends for the northern hemisphere. New Zealand data show little clear trend before 1940 and an increase of about 1 °C since then (Burrows, 1975; Salinger and Gunn, 1975; Trenberth, 1975). During the period 1967 to 1973 a large part of Australia experienced an increase in excess of 1 °C. There is also some evidence for a decrease of mean summer maximum temperatures over much of south-east Australia of about 1 °C between 1890 and 1940, with a possible slight increase in the period 1940 to 1950. Data from thirty Australian stations from 1957 to 1973 suggest that steadily decreasing temperatures are not a feature. Northern hemisphere evidence points to the largest cooling since 1940 having occurred in high latitudes, but observations from seven stations in Antarctica reveal no comparable downward trend (Fig. 2.3).

Australia and New Zealand may be anomalous in the context of the southern hemisphere as a whole. Streten (1977) reported that over the period 1958 to 1975 there was considerably more warming, or less cooling, in the Australian and New Zealand sector than elsewhere, contrary to trends in the rest of the southern hemisphere. Lamb (1975) suggested that trends in the 1960s and 1970s in at least part of the southern hemisphere, opposing northern hemisphere trends, may be related to southward displacement of the intertropical convergence zone, reducing net cross-equatorial transport of heat and increasing the release of latent heat south of the equator.

During the early twentieth century global warming period rainfall amounts increased generally in the zones of prevailing westerly winds in both hemispheres, from the west coasts far into continental interiors, but not in the lee of mountain ranges or towards the

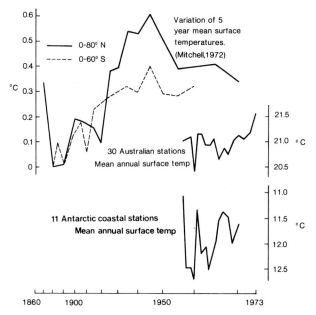

FIG. 2.3. AVERAGE SURFACE TEMPERATURE VARIATIONS
(After Tucker, 1975; reproduced with the permission of the Australian and New Zealand Association for the Advancement of Science).

east coasts. Subtropical anticyclones were more intensely developed than before and were in a zone of decreased rainfall. In the northern hemisphere south of the anticyclone belt the monsoon rains penetrated further northward into the southern fringe of the Sahara, and were more dependable than before in northern India.

Since the 1940s the vigorous zonal circulation seems to have been replaced by more meridional patterns, with smaller-scale systems effecting smaller-scale moisture transport. Equatorial rains have had a more restricted seasonal migration and their dominant position seems to have moved further south than before. This restricted seasonal migration has apparently led to rises in the East African lakes, and places in latitudes 10–20° N and 12–20° S have experienced droughts or monsoon failures repeatedly over a number of years (Lamb, 1975).

Both hemispheres have recently become more prone to blocking situations, with the migratory highs and lows tending to slow down and become more or less stationary in certain locations for periods of several days at a time. Since the controlling high pressure systems in such periods tend to occupy different longitudes in different years, places in middle latitudes have become liable to long spells of very dry or very wet weather, or with prolonged southerly or northerly winds, with abrupt changes from one extreme to the other.

Rainfall deficiency seems to have increased since about 1970 in extensive zones in both hemispheres between latitudes 10° and 30° (Fig. 2.4). A nearly continuous zone of increased rainfall developed in lower middle latitudes across the USA, the southern Mediterranean, Spain and Portugal. Several regions in high northern latitudes had substantial increases in rain and snow. In parts of the Canadian Arctic and Siberia, for example, precipitation totals for the early 1970s were up to 130 per cent of the 1931–60 averages, with up to 170 per cent of previous averages in Spitsbergen and parts of north-east Siberia. The 1970–72 rainfall averages in those parts of the temperate zone which became drier were mostly 85 to 90 per cent of 1931–60 normals, locally getting as low as 70 per cent. In the drought-afflicted zone of northern Africa, however, figures ranged down to 40 per cent and were under 25 per cent in the Cape Verde Islands (Lamb, 1974).

These rainfall fluctuations seem to constitute a phenomenon of global extent, characterized by a narrow zone of increased rainfall near the equator and some approach to symmetrical adjustments in both hemispheres on either side. The greatest recent reductions of rainfall have been near the tropics, with perhaps the most general increases in high latitudes.

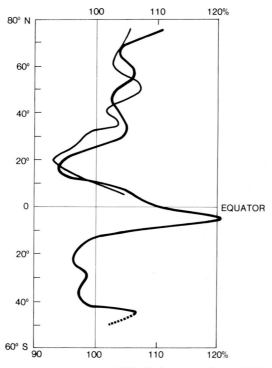

FIG. 2.4. DEPARTURES OF MEAN YEARLY RAINFALL 1970–72 (THICK LINE) AND 1960–69 (THIN LINE) FROM THE 1931–60 AVERAGE BY LATITUDE ZONES
(After Lamb, 1974; reproduced with the permission of Professor H. Lamb).

The few station records dating back to the 1800s suggest that much of eastern Australia was wetter in the nineteenth century than in the first half of the present century, but lack of data makes it difficult to extrapolate this pattern to the rest of the southern hemisphere. Shifts in climatic belts about the turn of the century had obvious consequences for agriculture and partly account for the failure of denser settlement in much of north-eastern South Australia in the 1890s (Fig. 2.5). At about this time there was a large decrease in summer rain and a smaller increase in winter rain. In the drier inland areas a decrease in annual rainfall occurred of up to 30 per cent. At many stations the change was quite sudden, occurring about 1893 to 1897. There seems to have been a return to higher annual rainfalls over much of eastern Australia around 1945 to 1946 (Fig. 2.6). Annual rainfall

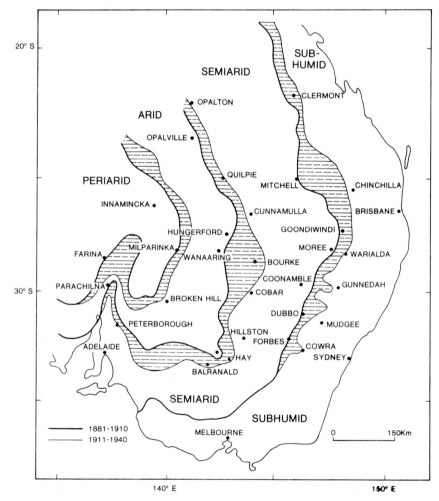

FIG. 2.5. SHIFTS IN THE CLIMATIC BELTS OF EASTERN AUSTRALIA, 1881–1910 TO 1911–40
(After Gentilli, 1971a; reproduced with the permission of Elsevier Scientific Publishing Co.)

has since increased by 10 to 20 per cent over a large area of south-east Australia. The increase along the coast of New South Wales and southern Queensland occurred mainly in summer and winter, while in inland New South Wales and Victoria it was in all seasons except winter. A decrease of about 10 per cent in annual rainfall in south-west Australia has been concentrated in winter (Australian Academy of Science, 1976; Cornish, 1977).

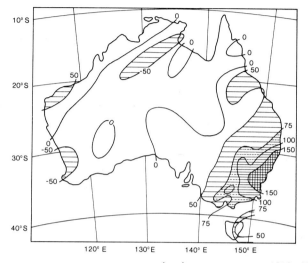

FIG. 2.6. CHANGES IN AUSTRALIAN MEAN YEARLY RAINFALL (mm) BETWEEN PERIODS 1913–45 AND 1946–74 (After Pittock, 1975; reproduced with the permission of the Australian and New Zealand Association for the Advancement of Science).

These examples from Australia provide good illustrations of the great variability that is possible within overall trends. It is interesting to note the estimate of Pittock (1975) that the area of significant (95 per cent level) rainfall change in Australia (1941–74 means compared with 1913–40 means) is about what would be expected from a random data set. Much of the observed variations in rainfall can be accounted for more or less as a random fluctuation, although there is no doubt that in some districts locally significant climatic changes have occurred, with important economic and social consequences.

Lamb (1974) has seen some parallelism between the course of climatic fluctuations in this century and those of the sixteenth and eighteenth centuries. The presently observed pattern in middle latitudes may imply that a very awkward kind of year-to-year variability is developing. Opposite extremes of warmth and cold, wet and dry have occurred in different sectors of the same zone. The same season at one place may tend to opposite extremes in different years, as blocking centres shift.

There appear to be common characteristics between the cooling trend since the 1940s and most earlier global climatic episodes. It is possible that all climatic cooling episodes are alike in being marked by declining strength of the mid-latitude westerlies and increased meridional circulation. All major episodes of this kind would therefore have some appearance of a glacial-onset regime. Whether a lasting increase of glaciation and permanent shift of the climatic belts results from any one such episode must depend critically on the solar radiation available during the recovery phases of the short-term fluctuations. It is possible that the influence of atmospheric pollutants might be just enough to swing the balance one way or the other.

2.3 AIR POLLUTION

Many forms of air pollution occur naturally. These include emissions from volcanoes and swamps, windblown dust, salt spray, vapours from leaves and from rotting material in the natural environment, and pollen from plants. Such pollutants can be considered to be accommodated within the global ecology. Natural sources of atmospheric pollutants are distributed over the entire globe, but man-made pollutants are injected at certain scattered spots, mainly in the industrialized continents. They include spray-can gases, aircraft, automobile, industrial, and domestic emissions from the use of fossil fuels, waste heat from numerous industrial and domestic sources, and noise.

In this section consideration is given to those man-made atmospheric pollutants that are considered to have possible climatic effects although other possible effects are examined in Chapter 3. Atmospheric pollutants contribute to atmospheric variability in two basic ways. Firstly, they change the composition of the atmosphere through time, and secondly they may be causes of climatic variability at different scales.

The residence times of pollutants in the atmosphere vary considerably, depending upon the nature of the pollutant itself, upon the way the emission has taken place, on meteorological factors, and on sink mechanisms. Usual sinks in the troposphere are chemical reactions, precipitation and gravitational fallout. Residence times in the stratosphere are usually considerably longer (perhaps 100 times) than those in the troposphere, because of the absence of effective sinks. Therefore, the stratosphere in particular seems to become a reservoir for man-made aerosols, as well as for volcanic particles.

Pollutants are transported by the wind and simultaneously diffused into larger volumes. During transport they are subject to photolysis, oxidation and reaction with other gases. Table 2.3 lists those man-made atmospheric pollutants that are considered to have potential climatic effects.

2.3.1 Carbon dioxide

Since the beginning of the Industrial Revolution man has been burning fossil fuels, and in the process releasing carbon dioxide, water vapour, sulphur dioxide, various nitrogen oxides, ash, dust, smoke, and heat to the atmosphere. About half of the carbon dioxide so produced is still in the atmosphere; the other half has been dissolved in the oceans or has gone into the earth's biomass (Budyko and Karol, 1975; Kellogg, 1977).

Our chief concern with this changing component of the atmosphere is its effect on the heat balance, since carbon dioxide is virtually transparent to solar radiation but absorbs terrestrial radiation in several wavebands. Additional carbon dioxide enhances absorption of terrestrial radiation, thereby warming the lower atmosphere, which reradiates back downwards, thereby warming the surface.

Table 2.3
MAN-MADE ATMOSPHERIC POLLUTANTS AND POTENTIAL EFFECTS
(partly after Almqvist, 1974; Knox and MacCracken, 1976)

Pollutant	Main anthropogenic sources	Potential atmospheric effects
Carbon dioxide	Combustion of fossil fuels	Increased temperatures
Fluorocarbons	Aerosol cans, Refrigeration systems	Reduction of stratospheric ozone, disturbances of radiation balance
Nitrogen oxides	High-flying aircraft, Combustion, Fertilizers	Reduction of stratospheric ozone, disturbance of radiation balance, particle formation
Sulphur compounds	Combustion of fossil fuels	Particle formation, precipitation chemistry
Waste heat	Industry, Space heating and cooling, Power stations	Increased temperatures, precipitation modification, circulation modification
Water vapour	Combustion, Cooling towers	Disturbance of radiation balance, particle formation
Ammonia	Waste treatment	Particle formation
Hydrocarbons	Combustion, Chemical processes	Particle formation
Methane	Chemical processes	Stratospheric water vapour and ozone concentrations
Peroxyacetylnitrates	Oxidation of olefins from factory effluents	Suspected but unknown
Methyl bromide	Agricultural fumigation	Destruction of stratospheric ozone
Krypton-85	Nuclear fuel reprocessing and power plants	Modification of electric field, possible modification of precipitation processes
Aerosols	Combustion, Agriculture	Temperature change, precipitation modification

Recent measurements show a continuing increase in atmospheric carbon dioxide concentration (Fig. 2.7). The present background concentration is approximately uniform over the globe at about 330 p.p.m. which is about 10 per cent more than estimates of 290–300 p.p.m. for the preindustrial levels, thought to have prevailed for 10 000 years or more. Observations in Hawaii and at the South Pole have shown an average concentration rise of 14 p.p.m., 1958 to 1974 (Keeling et al., 1976a, b). The annual rate of increase rose from about 0·7 p.p.m. per year in the late 1950s to nearly 1·3 p.p.m. per year in 1973–1974. The reason for the recent change is uncertain, but it may be related through energy consumption to the so-called 'energy crisis'.

Carbon dioxide concentrations measured at 3 to 5 km over south-east Australia since 1972 show both an upward trend and a seasonal variation (Fig. 2.8). The data indicate that the increases in annual mean concentrations from 1972 to 1973 and from 1974 to 1975 were about twice those which occurred from 1973 to 1974 and from 1975 to 1976. It has been speculated that variability in the rate of carbon dioxide increase is related to the

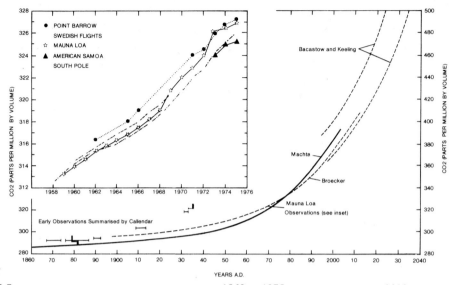

FIG. 2.7. THE RECORD OF ATMOSPHERIC CARBON DIOXIDE, 1860 TO 1975, WITH PROJECTIONS TO 2030
(After Kellogg, 1977; reproduced with the permission of the World Meteorological Organization).

Southern Oscillation, which is a fluctuation of atmospheric circulation with a period of two to three years (Bacastow, 1976). It involves pressures, winds, rainfall, air temperatures and sea surface temperatures in many parts of the world, but especially in the tropical zones of the Pacific and Indian Oceans. The ability of the ocean to absorb carbon dioxide depends on the strength of the atmospheric circulation, of which the state of the Southern Oscillation is a measure. Correlation between eastern tropical Pacific warmings and a Southern Oscillation index points to a possible mechanism. Variations in the former may explain year-to-year variability in oceanic uptake of carbon dioxide, as well as the phase

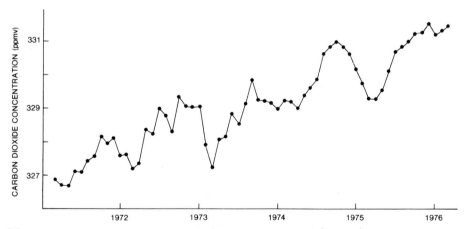

FIG. 2.8. MONTHLY MEAN CARBON DIOXIDE CONCENTRATIONS OVER AUSTRALIA (p.p.m.v.)
(After Pearman, 1977; reproduced with the permission of the Clean Air Society of Australia and New Zealand).

between deviations from the average annual variation in carbon dioxide variations observed at Hawaii, the South Pole and in Australia (Pearman, 1977).

Most models predict a doubling of carbon dioxide concentration in about 50 to 75 years. There are considerable uncertainties, however, in model representations of the carbon cycle. One area of uncertainty, for example, concerns the estimates of natural flux rates of carbon dioxide between the atmosphere, biosphere and hydrosphere. It is not at all unlikely that currently observed increases in carbon dioxide levels are due at least in part to a slight variation in source (and/or sink) strengths of the natural cycle.

Rotty (reported in Pearman, 1977) has provided updated estimates of fossil fuel carbon dioxide production, together with predictions of future use (Fig. 2.9). The 1976 carbon

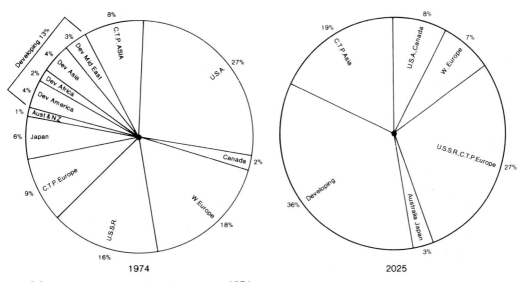

FIG. 2.9. GLOBAL CARBON DIOXIDE PRODUCTION IN 1974
(After Pearman, 1977; reproduced with the permission of the Clean Air Society of Australia and New Zealand).

dioxide production was estimated at $5 \cdot 05 \times 10^{15}$ g C, with a predicted release of 26×10^{15} $gC.a^{-1}$ in 50 years time. Various researchers have prepared scenarios using present estimates of fossil fuel reserves and assuming that the rate of increase of consumption will decrease, but with different assumptions concerning values for uptake by the oceans (Table 2.4). Results indicate an increase in atmospheric carbon dioxide concentrations of between four and eight times present levels in the next 200 years, with a slow decline thereafter. It is interesting to note that with predicted population growth rates and improved living standards in developing countries, it might be expected that such countries will consume an increasing proportion of total fossil fuels. This points to the importance of energy policies in countries other than those presently consuming most of the fossil fuels.

If there is uncertainty in the prediction of carbon dioxide trends, then predictions of the resulting climatic effects are even more uncertain. Schneider (1975) reviewed various climate models used to predict change resulting from increasing atmospheric carbon dioxide. He concluded that global surface temperature sensitivity to a doubling of

Table 2.4
RECONSTRUCTION AND PREDICTION OF ATMOSPHERIC CARBON DIOXIDE CONTENT
BASED ON FUEL CONSUMPTION DATA
(after Broecker, 1975)

| Year | Chemical fuel CO_2 ($\times 10^{16}$ g) | Excess atmospheric CO_2[a] | | | CO_2 content of atmosphere (p.p.m.) | Global temperature increase (°C)[c] |
		($\times 10^{16}$ g)	%	(p.p.m.)[b]		
1900	3·8	1·9	0·9	2	295	0·02
1910	6·3	3·1	1·4	4	297	0·04
1920	9·7	4·8	2·2	6	299	0·07
1930	13·6	6·8	3·1	9	302	0·09
1940	17·9	8·9	4·1	12	305	0·11
1950	23·3	11·6	5·3	16	309	0·15
1960	31·2	15·6	7·2	21	314	0·21
1970	44·0	22·0	10·2	29	322	0·29
1980[d]	63	31	14	42	335	0·42
1990	88	44	20	58	351	0·58
2000	121	60	28	80	373	0·80
2010	167	83	38	110	403	1·10

[a] assuming 50% of CO_2 produced by burning fuel remains in atmosphere
[b] preindustrial atmospheric partial pressure of CO_2 assumed to be 293 p.p.m.
[c] assuming 0·3 °C global temperature increase for each 10% rise in atmospheric CO_2 content
[d] post 1972 growth rate taken as 3% per year

carbon dioxide to about 600 p.p.m. can be given at 1·5 to 3·0 °C. He also pointed out that the response could be several times higher at high altitudes and that feedback mechanisms could modify estimates several-fold. Possible feedback mechanisms such as increased cloudiness would reduce incoming solar radiation and so tend to counteract warming, but increased sea surface temperatures could mean reduced carbon dioxide uptake in the oceans and hence enhanced warming.

The general circulation model of Manabe and Wetherald (1975) predicts a global surface temperature increase of 2·9 °C for a doubling of carbon dioxide (Table 2.5). The model also predicts that the activity of the hydrologic cycle will be increased by 7 per cent,

Table 2.5
CALCULATED HEAT BALANCE CHANGES (%) FOR AN ATMOSPHERE WITH PRESENT (1975)
CARBON DIOXIDE LEVELS AND TWICE THESE
(after Bach, 1976; data from Manabe and Wetherald, 1975)

Fluxes	Change (%)
Earth Surface	
Net short-wave radiation	+0·5
Net long-wave radiation	−6·6
Net all-wave radiation	+3·3
Latent heat	+6·9
Sensible heat	−7·7
Top of Atmosphere	
Net short-wave radiation	+0·9
Net long-wave radiation	+0·9

but it allows for no feedback to increase, or change the character of, cloudiness. The possibility of both positive and negative feedbacks in the real atmosphere casts doubt on the validity of such models. On the other hand, analyses by Cess (1976) suggest that cloud amount is not a significant feedback mechanism, and other recent experiments have supported this conclusion.

The main sink for carbon dioxide in the long run will be the oceans (Bolin, 1975; Keeling, 1977; Oeschger et al., 1975), since, contrary to increasing, the forests of the world are quite possibly being cut down faster than they can grow. The oceans contain about 60 times more carbon dioxide than the atmosphere, but for them to reach a new equilibrium with a larger atmospheric content there has to be an exchange between the upper levels of the oceans (100 to 1000 m) and the deep ocean water (Kellogg, 1977). It has been estimated that this process takes at least 1000 years, and Keeling (1977) estimates a decay time for atmospheric carbon dioxide of 1500 years. Thus, even if we could stop releasing carbon dioxide from fossil fuel in say the next century, we would still find the incremental carbon dioxide lingering in the atmosphere at a very slowly diminishing concentration for many centuries.

There is still considerable indecision about the relative importance of fossil fuel input and deforestation in the context of the carbon cycle. Pearman and Garratt (1972) calculated that global trends in deforestation could be a source of atmospheric carbon dioxide of the same magnitude as fossil fuel input. Woodwell and Houghton (1976) concluded that net input from biota could range from 50 to 400 per cent of the fossil fuel input.

It is recognized that changes of a few degrees in global mean temperature would have a significant effect on society in general and on agriculture in particular (e.g. Pittock, 1972). A 2 °C warming would return Earth to a condition similar to that which existed during the climatic optimum 6000 years ago. In the long term such changes may actually be beneficial to some areas; but in the short term the rate of change suggested by the available data could impose great stress on society.

Increased carbon dioxide concentrations might also lead to other, less direct effects on climate. If the ocean surface layers were to absorb more carbon dioxide their acidity would increase gradually, leading to possible changes in biological productivity, which in turn could affect the ocean surface albedo. Other, indirect effects requiring further examination include the magnitude and time scale of the response of glaciers and polar ice to increasing temperatures. Consideration might also be given to the effects of variations of atmospheric temperatures upon concentrations of various trace gases. It has been suggested for example, that ozone concentrations would rise 2 to 5 per cent as a result of stratospheric cooling related to an average atmospheric temperature increase of several degrees (Pearman, 1977). Yet another possibility concerns the development of carbon dioxide concentrations so high that total atmospheric pressure would be affected (this may require increase by a factor of as much as 1000). Absorption bands could then broaden, opacity would increase and temperature may start to rise so rapidly that the process could run away. This does appear a remote possibility, even on a geological time scale (Rasool and Schneider, 1971).

2.3.2 Aerosols

Particles, commonly known as aerosols, are added to the atmosphere by industry, power generation, automobiles, space heating, agricultural practices and most activities of man. Most of the particles are ejected to the atmosphere over the large cities and industrial centres, where they are produced largely by a combination of fossil fuel burning (which produces soot and sulphur dioxide, the latter becoming sulphate particles after a short time), and the creation of particles from unburned hydrocarbons by photochemical reactions in the presence of solar ultra-violet radiation. Such secondary particles (sulphates and hydrocarbons) tend initially to be somewhat smaller in size than the directly produced smoke or soot particles. The majority of man-made particulates are oxidation products of gaseous emissions including sulphur dioxide and nitrogen oxides. Particles so formed consist of sulphuric acid, nitric acid or ammonia-neutralized salts thereof. Most small atmospheric particles are of ammonium sulphate or sulphuric acid (Barrie et al., 1976; Bigg, 1977).

There seems little doubt that since the turn of the century there has been an increase in the rate at which aerosols have been produced by mankind, particularly in the more industrialized countries (SMIC, 1971). A review of turbidity observations at several stations with long records (Machta and Telegades, 1974) includes eight Soviet stations whose trends show varying degrees of increase in total aerosol content. All but one are located in or near large cities which have grown in the last century. Similar increases in turbidity have been observed at Davos (Switzerland), Washington DC, Mexico City, and Jerusalem.

On the other hand, observations of total suspended particulates at 18 non-urban stations in the USA showed no increase from 1960 to 1972. Observations in Hawaii have similarly shown no long-term increases, except those that can be accounted for by large volcanic eruptions in the tropics (Kellogg et al., 1975). Dyer (1974) also reported no convincing evidence for a long-term increase in turbidity, although volcanic activity has frequently produced a short-term decrease in atmospheric transmission of solar radiation. Evidence from observations of electrical conductivity over the oceans suggests increases in aerosol content near the surface (decreased conductivity) in this century in the western North Atlantic, western North Pacific and the Indian Ocean close to India.

The overall picture seems to be that aerosol trends are upward in most of Europe and the Soviet Union, where there has been increased industrialization and where pollution abatement efforts lagged, at least before the 1970s. In the USA, the UK and some other countries where vigorous pollution abatement efforts were well under way in the early 1960s, reduction seems to have approximately kept pace with the increased industrial activity. Over the open oceans increases have apparently been limited to regions immediately downwind of large population concentrations.

All major sources of suspended particles are surface-based and hence, with the exception of volcanoes, they pollute the lower layers of the atmosphere (0–5 km). The average residence times of small particles are not known with certainty, although it is known that they depend on the range of particle sizes, the altitudes of injection and various other factors. In general, however, particles entering near the Earth's surface have the shortest residence times of up to about ten days, whereas those entering in the stratosphere at altitudes above 20 km may reside for a year or more. Near-surface sources are at least two or three orders of magnitude more prolific than stratospheric sources, both natural and man-made. The short residence times mean that the particles are unlikely to become thoroughly mixed like carbon dioxide and the greatest build-ups of suspended particulates are likely to occur within about 1000 km of major source regions (Barrie et al., 1976).

The world-wide aggregate source strength of particles has been estimated at about 3×10^9 t.a^{-1}, of which about $1 \cdot 6 \times 10^9$ t.a^{-1} consist of particles smaller than 5 μm diameter. Over cities atmospheric concentrations of particles may reach 100 000 cm^{-3} (Bigg, 1977). This compares with a background level of about 200 to 400 cm^{-3} in clear air over Antarctica or the remote parts of the oceans. Air over remote parts of continental Australia averages about 700 particles cm^{-3}. Particle levels over continental Australia are remarkably low, especially considering the relative absence of clouds, which normally help to mop up particles. Measurements near Lake Eyre have even given concentrations near the oceanic background level.

Table 2.6
PRODUCTS OF PORT KEMBLA STEELWORKS WITH
POSSIBLE METEOROLOGICAL SIGNIFICANCE
(after Ogden, 1969)

Sensible heat to air	$3 \cdot 4 \times 10^3$ MW
Sensible heat to sea	$0 \cdot 4 \times 10^3$ MW
Latent heat as WV	$0 \cdot 4 \times 10^3$ MW
Total heat	$4 \cdot 2 \times 10^3$ MW
Water evaporated	170 kg.s^{-1}
Condensation nuclei	10^{14} .s^{-1}
Freezing nuclei	10^8 .s^{-1}

Industrial centres put large amounts of sulphur dioxide into the atmosphere (Table 2.6). Mt. Isa in Queensland, for example, produces a great deal of sulphur dioxide, vast numbers of very small sulphuric acid particles and relatively few giant particles. The pollution plume can be detected easily up to at least 500 km from Mt. Isa, at which point it is about 300 km wide. The numbers of particles seem to increase downwind, which is surprising. This may be explained by the fact that in the presence of water vapour and under the influence of sunlight the particles are formed by sulphur dioxide changing to sulphuric acid. Near the source, when the particles are close together, the acid condenses on to already existing particles. As the plume becomes diluted by the atmosphere, new particles form instead. It is probable that the particle output from Mt. Isa trebles in a day's travel. Particle concentrations of five times the background level have been found downwind of Perth and Kwinana (Bigg, 1977). It seems reasonable to expect similar, or worse, situations in other, more industrialized parts of the world.

Aerosol particles can both scatter and absorb sunlight, and they also absorb and re-emit infra-red radiation. When a non-absorbing particle scatters solar radiation some of the scattered radiation will be lost to space and some will be directed downward. The result is less sunlight reaching the Earth and an increase in the net albedo of the earth-atmosphere system, which would cause a net cooling. Some global climatic models have predicted a cooling of between 1 and 3 °C for a doubling of the present aerosol concentrations.

On the other hand, absorption of solar radiation heats the particle and the air around it, the effect of which is to reduce the net albedo (Kellogg, 1977). The magnitude and sign of net albedo, hence temperature, changes caused by lower atmosphere aerosols depend on several uncertainties: the composition and concentration of aerosols; the size distributions of the particles; the strong humidity dependence of the degree of absorption of terrestrial radiation (longwave absorption generally increases with humidity); and the existence of feedback mechanisms, such as cloud cover, which have been largely ignored or only crudely accounted for in existing models (Barrie et al., 1976).

The distinction between warming and cooling attributable to aerosols takes into account the ratio of the particle absorption (a) to its backscatter (b), and also the albedo of the underlying surface. When aerosols of a given a : b ratio are over a dark surface, such as the ocean, they are more likely to increase the net albedo than when they are over a light surface, such as a snowfield, or over land generally.

There has been a widely held belief that anthropogenic aerosols promote cooling. This line of thought seems to be based on an assumption that aerosols would be evenly spread around the globe. Recently, however, it has become more generally accepted that most anthropogenic aerosols exist over land, near where they are formed, and that they are sufficiently absorbing to reduce the albedo and so contribute to surface warming (Kellogg, et al., 1975; Bigg, 1977).

There is still some uncertainty, therefore, as to whether aerosols are cooling or warming the Earth's surface, or doing neither. Regular measurements are needed of aerosol optical characteristics, as well as improved three-dimensional numerical models of the atmosphere. Anthropogenic aerosols may alter the Earth's radiation balance indirectly by interfering with cloud processes. For a given liquid water content the reflectivity of a cloud is controlled by the abundance of cloud droplets, which in turn depends on the number of cloud condensation nuclei. Most man-made aerosols, particularly sulphate and nitrate compounds, are good cloud condensation nuclei, but their effects on cloud cover are not known. Mitchell (1971) suggested that aerosol climatic effects are likely to include a slight decrease of cloudiness and precipitation.

2.3.3 Fluorocarbons (chlorofluoromethanes)

A World Meteorological Organization statement on anthropogenic modification of the ozone layer (WMO, 1976) concluded that evidence supports the view that a continued release of chlorofluoromethanes into the atmosphere may lead to a significant reduction in stratospheric ozone. There is still great uncertainty, however, about the size of the impact of fluorocarbons on ozone (Anon, 1977). So far, estimates suggest that ozone concentrations may have dropped by about 0·5 to 1·0 per cent, an amount that would be obscured by much larger natural fluctuations. The WMO statement points to a possible eventual steady-state 10 per cent average ozone depletion. On the other hand the picture is confused by suggestions such as those of Parry (1977) that the ozone concentration might be rising and that chlorines do not necessarily cause depletion. It is also confused by recent reports (December, 1978) from research in Hawaii that volcanic eruptions may produce fluorocarbons naturally.

Aerosol spray cans account for about 75 per cent of the global emissions of fluorocarbon gases (freons). The rest is divided between air conditioning and refrigeration on one hand, and foam plastics (polyurethane, polystyrene) on the other. In 1975 about 3000 million cans ejected more than 500 000 t of fluorocarbons, and the total output of freons (mainly Freon-11 and Freon-12) was nearly 700 000 t. World-wide production and release of F-11 and F-12 grew by about 10 per cent per annum up to 1974, but fell in 1975 and 1976 by about 15 per cent. It is uncertain how much of the fall in output was due to concern about the hazards of fluorocarbons, how much to the world economic slump and the economic attractiveness of competing products.

Current monitoring indicates that the atmospheric concentrations of fluorocarbons are rising rapidly, with data for F-11 showing an annual increase of from 13 to 28 per cent. Measurements in the region above Australia showed a 19 per cent increase between April 1976 and March 1977.

Fluorocarbons seem to spread rapidly through the troposphere so that concentrations in the southern hemisphere are only 5–10 per cent lower than in the northern hemisphere, despite the fact that 95 per cent of emissions occur in the latter. Southern hemisphere concentrations in 1976 were about 120 p.p.m.m. by volume for F-11 and 200 p.p.m.m. for F-12.

Stratospheric concentrations are well below those in the troposphere, probably because of the slow rate of movement of air from the troposphere to the stratosphere. Fluorocarbons are finally broken down by ultra-violet radiation in the stratosphere. The average freon molecule probably takes several decades to reach above 25 km in the stratosphere, where the breakdown occurs. Fluorocarbons themselves are unusually chemically inert, so unlike many pollutants they are not quickly broken down or removed. It is the products of the eventual breakdown that take part in ozone-destroying reactions, so it is likely to be several decades before the fluorocarbons released now have any impact on ozone levels.

Stratospheric ozone concentrations fluctuate with natural changes in the rates of continuous production and destruction. In any year the maximum concentrations in spring can be half as high again as autumn minima, and smaller changes in average concentrations occur from year to year. According to Kulkarni (1976) changes in lower stratospheric circulation are sufficient to explain any observed trends in stratospheric ozone concentrations, and there seems no need to invoke chemical reactions for the destruction of ozone. Nevertheless, the weight of opinion appears to support the view that while man's activity is unlikely to change ozone production rates, the addition of chemicals such as F-11 and F-12 may indirectly increase the rate of destruction and so reduce average concentrations.

The main natural destroyers of ozone are oxides of nitrogen, which convert perhaps 60–70 per cent of ozone produced back to oxygen. Oxygen atoms, hydrogen atoms and various combinations of hydrogen and oxygen are also ozone destroyers. Chlorine atoms and a combination of chlorine and oxygen are minor natural destroyers, but they are also the destructive agents contributed by F-11 and F-12. The impact of chlorine released from the fluorocarbons is multiplied because chlorine can destroy thousands of ozone molecules through catalytic reaction cycles. Destruction stops only when other reactions occur that incorporate the chlorine in compounds that do not react with ozone.

Reduction of stratospheric ozone concentrations, and the accompanying shift in the distribution of ozone towards lower altitudes, could change the temperature distribution in the stratosphere, with the effects flowing on to the lower atmosphere (Fig. 2.10). More ultra-violet and visible radiation would reach the ground, tending to warm the lower atmosphere and the Earth's surface. At the same time reduced absorption of ultra-violet radiation in the stratosphere would reduce heating there. This in turn would tend to promote surface cooling, as less thermal radiation would be emitted from the stratosphere to the ground. These effects, one of warming the other of cooling, would be complicated by the changed distribution of ozone in the stratosphere. The net outcome of fluorocarbon-induced ozone reductions is, therefore, unpredictable.

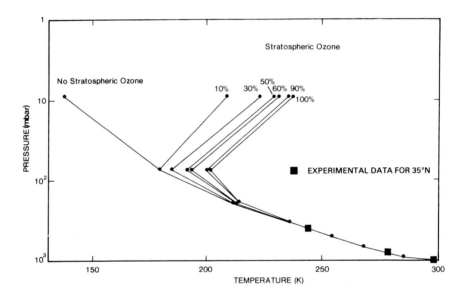

FIG. 2.10. CALCULATED STEADY STATE TEMPERATURE PROFILES, AS A FUNCTION OF PRESSURE (mbar) FOR THE PRESENT ABUNDANCE OF STRATOSPHERIC OZONE AND FOR 90, 60, 50, 30, 10 AND ~0 PER CENT OF THOSE VALUES. MEAN GLOBAL SURFACE ALBEDO OF 0·1
(After Reck, 1976; reproduced from *Science*, 192, p. 557, with the permission of the American Association for the Advancement of Science and Dr R. Reck; copyright 1976 by the American Association for the Advancement of Science).

If releases of F-11 and F-12 continue at about present rates, concentrations might reach about 700 p.p.m.m. and 1900 p.p.m.m. respectively by next century. The direct heat-trapping effects of fluorocarbons have been estimated on this basis as likely to average 0·5 °C at the surface. Chlorofluoromethanes may significantly enhance the carbon dioxide greenhouse effect, since they are particularly strong absorbers in the 8–13 μm band, where water vapour and carbon dioxide are relatively poor absorbers (Barrie et al., 1976). As with heating effects induced by other pollutants, any global temperature rise would be made up of much bigger rises in some parts of the world than in others. This would induce other climatic changes, notably increased rainfall in some areas. Effects on agriculture would probably depend much more strongly on regional and local changes than on global average changes.

2.3.4 Thermal pollution

Total world energy production in 1977 was about 10^4 GW, compared with the rate of absorption of solar energy at the Earth's surface of 8×10^7 GW (Kellogg, 1977). Per capita energy use, currently about $2 \cdot 5$ kW, is growing, possibly at more than 5 per cent per year. Energy production growing at such a rate could lead to a situation where man's energy input to the atmosphere might become significant since estimates suggest that a variation of only 1 per cent in solar output could be expected to have significant climatic effects.

The world's atmosphere constitutes a very large and energetic system compared with which man's activities are relatively minor. A large low pressure system, for example, generates and dissipates kinetic energy at a rate of about 100 W.m^{-2}, compared with which the total power consumption in the UK is only about $1 \cdot 5$ W.m^{-2}. Nevertheless the atmospheric system may be sensitive to small variations from its natural state, so the possibility of human effects should not be ignored. Release of thermal energy to the environment is increasing, and also the means for abolishing other pollutants are almost invariably energy-intensive, hence producing further thermal pollution.

Local and regional concentrations of heat release might produce heat sources to upset the atmospheric circulation much sooner than global effects become significant. Principal emissions are from industrial processes, space heating and cooling, and power stations. Waste heat may be emitted directly to the atmosphere or indirectly via cooling ponds, rivers and lakes. On a local scale man-made and solar heat inputs can be of the same order of magnitude. In very heavily populated cities the man-made heat input may even exceed solar heat input, particularly in winter (Barrie et al., 1976).

It is possible that release of heat from large conurbations and proposed power parks will pose problems, although the effects of such regional sources on global weather patterns have not yet been determined. Hanna and Gifford (1975) have suggested that heat emissions from a 4×10^4 MW nuclear power park would initiate and augment convective clouds and precipitation over the source area, cause slight increases of ground fog within 100 km of the cooling towers and a significant persistence of water vapour plumes.

A different perspective can be introduced by considering the behaviour of the oceans. From year-to-year the oceans vary in temperature by about 1 °C. A temporarily warm ocean provides a regional heat source which is responsible for at least some of our year-to-year climatic variations. To have a similar effect man's heat output would have to match that of the ocean over a corresponding area. Sawyer (1975) concluded that this would require an output of about 50 W.m^{-2} over an area comparable in size to Western Europe. Present industrial and domestic heat output is about 1 W.m^{-2}, so a fifty-fold increase would be needed before a climatic effect resulted comparable to the natural year-to-year variations.

Barrie et al. (1976) made a similar comparison for the area of the Great Lakes. In winter the open Great Lakes add about 10^7 MW of heat to the atmosphere from an area of 25×10^4 km^2. This input is known to intensify low pressure systems and associated weather patterns. An equivalent man-made effect would require about 10^9 average Americans and their associated industry in such an area.

2.3.5 Other pollutants

There has been considerable research into the possible effects on the ozone layer of nitrogen oxides emitted by supersonic aircraft. There are many uncertainties in both measurements and theory, but the likelihood is that the risk to the ozone layer is much smaller than from fluorocarbons (Anon, 1977).

A recent theory suggests that increased use of agricultural fertilizers and/or of nitrogen fixing vegetation might affect the natural nitrogen cycle and result in increased release of nitrous oxide from the surface into the atmosphere. Emissions of nitrous oxide as a normal part of the nitrogen cycle, by microbiological reactions in soil, are the main source of ozone-destroying nitrogen oxides in the stratosphere and hence play a role in determining the natural abundance of stratospheric ozone. Present fertilizer use is believed to add no more than a few per cent to the natural emissions of nitrous oxide.

Nitric oxide is a major pollutant from combustion. It is a precursor of nitrogen dioxide, the trigger chemical in smog formation. Natural sources may, however, produce at least seven times as much nitric oxide and nitrogen dioxide as do man-made sources (Almqvist, 1974).

Stratospheric water vapour content seems to have been increasing over the last 15–20 years, possibly due to volcanic activity but possibly also due to supersonic aircraft exhausts. Water vapour in the stratosphere plays an important role in ozone photochemistry, in the formation of aerosols, and in the radiation balance of the Earth.

Hydrocarbons are produced mainly in the use and processing of petroleum. Little is known about their natural cycles, but they are known to be important in urban smog production, together with nitrogen oxides and solar radiation. They also influence aerosol production.

Ammonia comes principally from biological decay, but is also produced in waste treatment. It is important for producing aerosols in interaction with sulphur dioxide. Such aerosols make an important contribution to the stratospheric aerosol layer and may therefore affect the global heat balance. Methane is mostly naturally produced, but some comes from man-made chemical processes. It influences water vapour and ozone concentrations in the stratosphere, and hence may have some effect on solar radiation receipt. Compounds such as methyl bromide, used in agricultural fumigation, could contribute to stratospheric ozone destruction (Anon, 1977).

Boeck et al. (1975) have warned that increasing emissions of krypton-85 from nuclear power generating stations could raise the electrical conductivity of the lower atmosphere by 15 per cent in 50 years. Consequent changes in thundercloud electrification could alter precipitation processes and thereby modify climate. As with many other aspects of possible climatic effects resulting from man-made air pollution, however, this still has to be verified.

2.3.6 The future

Man has interfered with the traditional pathways of trace substances through the biosphere, particularly carbon, sulphur and nitrogen, with possible implications for climatic change (Barrie et al., 1976). Some waste chemicals have been and are being discharged in proportions sufficient to alter the natural composition and radiative properties of the atmosphere. Detection of imbalance of carbon dioxide, oxygen and water vapour, for example, implies imbalance in any natural cycle in the biosphere (Almqvist, 1974).

We still need to know, however, what kind and amount of environmental stress derived from human activities might be tolerated by the climatic system before it responds with an important climatic change. We also need to know more about the possible dimensions of natural climatic variability, but are still a long way from understanding the complex interactions of the many physical processes that determine the evolution of climate. It is still difficult to trace cause and effect linkages in the production of climates (Fig. 2.11). Many feedback influences have unknown effects, some may operate in one direction, some in the opposite direction, some may dampen out potential climatic effects of increased pollution, and others may accentuate the effects (Schneider, 1974b).

The concensus seems to be that the impact of an increased anthropogenic aerosol loading cannot be assessed reliably, but that the net effect will probably be small. Continuation of the present energy growth rate may mean that within one generation the production of waste heat will reach an amount that has been shown in natural processes to cause climatic changes (Budyko and Karol, 1975). The effects of carbon dioxide on climate may cause the biggest problems. These effects may be increasing because of the slow exchange of carbon dioxide with the deep ocean layer. It is not so much the total fossil fuel consumption as the speed of consumption that is responsible for the diminishing effectiveness of the oceans as absorbers. The critical nature of the carbon dioxide problem has been placed in perspective by the suggestion that the time required for a new energy production system to gain a 5 per cent share of the total market is about 60 years in the USA, and longer for the rest of the world. This is comparable to the predicted time for doubling of preindustrial carbon dioxide levels, assuming continued fossil fuel use.

Avoidance of carbon dioxide accumulation in the atmosphere is a problem. If the oceans were freely mixed they would be capable of taking up perhaps 80–90 per cent of the excess carbon dioxide. One possibility might be the trapping of carbon dioxide from industrial exhaust systems and piping it directly to the deep oceans. Alternatively, the oceans could be fertilized with nutrients that at present limit productivity, and thus increase sedimentation of carbon into the deep ocean. In both cases, however, cost would probably be prohibitive, additional use of fuel would be required, and there may be associated undesirable environmental effects (Pearman, 1977).

Reactions in the stratosphere are of particular importance to global climate. The US National Research Council has estimated that if world-wide F-11 and F-12 production continues at the 1973 rates, there is a 95 per cent chance that stratospheric ozone amounts will be reduced by 2–20 per cent, most likely about 7 per cent. Changes in the stratosphere take place very slowly, so it would take 40–50 years for only half of the ultimate ozone reduction to occur (Anon, 1977). Substances from chlorofluoromethanes have been found to destroy ozone six times more efficiently than nitrogen oxides. It has also been found that bromium, unlike chlorine compounds or nitrogen oxides, can destroy ozone at night and at lower altitudes. The possibility cannot be ruled out that other such substances will be identified in the future.

The trend away from fluorocarbons has already started, so that, for example, about 50

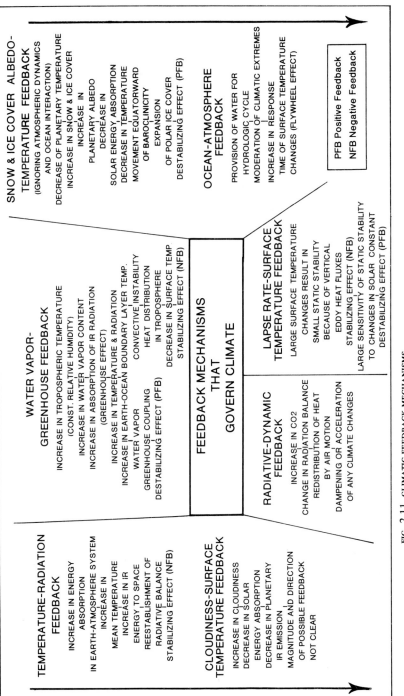

FIG. 2.11. CLIMATIC FEEDBACK MECHANISMS
(After Bach, 1976; reproduced with the permission of Ferd. Dummlers Verlag).

per cent of the aerosol spray cans sold in Australia use hydrocarbon propellants. The US Food and Drug Administration and Environmental Protection Agency required all cans using fluorocarbons to be labelled to that effect from 26 October 1977. There were also plans for phasing out non-essential uses of fluorocarbons in spray cans, to take effect from November 1978, with a probable phasing out period of about six months. Canada is following the lead and Sweden has already banned fluorocarbons in spray cans.

The potentially serious threat from pollutants is that because they build up to measurable levels only over long periods of time they involve large capital investments and are thus hard to change once initiated. An argument might be made for a reduction in fossil fuel consumption. This might be achieved by a reduction in the effective amount of energy consumed per person, but is, of course, accompanied by implied changes in the attitudes of society and encouragement of reduced population growth.

The net impact of human activities on climate is difficult to determine or predict. It is particularly difficult to isolate changes caused by human activity from those which might occur naturally anyway because of external factors, such as varying earth–sun relationships. Nevertheless, there is considerable speculation that climatic changes could result from man's activities, and that natural changes might be accelerated or slowed down by inadvertent modification of the atmosphere. Climatic changes could necessitate major agricultural and pastoral adjustments if total food production is not to fall (Schneider, 1977).

In summary, we know that climates change and that atmospheric pollution is increasing. We do not know with any degree of certainty the nature of the linkages, if any, and in the cases of some pollutants we are even unsure about the direction of their effects, if any.

2.4 HAZARDS AND EXTREMES

Variability is an inherent characteristic of climate. The discussion so far has centred upon possible variability over periods of many years, with changes of a gradual rather than violent nature. Even at the level of seasonal or diurnal variability, the changes tend to be within the range to which man can adjust relatively easily, although climatic change is probably accompanied by a different set of extreme and hazardous conditions, which may make adjustment difficult. On the other hand violent weather variations, such as tropical cyclones, tornadoes, hailstorms and lightning frequently result in catastrophic losses of property and income, as well as lives. Less violent, but nevertheless equally hazardous, atmospheric extremes include droughts, heatwaves, snow, fog, and frost. In addition there is the obvious atmospheric contribution to events such as floods.

Extreme events, particularly the most intense storms, are often of the greatest importance to man, for example in relation to erosion, river channel design and water supply. The occurrence of climatic extremes is of great importance in matters such as the design of structures in areas prone to tropical cyclones, or in the design of water storage dams in areas subject to long dry spells.

Environmental analysis tends to focus on situations which are most common and which are most likely to apply to the majority of cases considered. Sometimes, however, the abnormal, the infrequent, and the extreme can have major effects. The response of a society to environmental conditions is geared primarily to its perception of the 'normal' event, so the impact of an extreme of considerable magnitude can introduce a new set of conditions related to a different environmental balance. Many observable patterns in the environment may be explained more by extreme events than by the more frequent occurrence. In some cases an extreme frost, a severe snowstorm, or a major drought can be the ecological determinant of a vegetation association, rather than the more persistent characteristics of the habitat.

In the case of atmospheric phenomena such as rainfall it is important to recognize that annual figures do mask extreme events, which may in fact be responsible for most of the annual total at some locations. In arid areas in particular the intensity and irregularity of storms make annual averages meaningless. At Alice Springs in central Australia, for example, the annual mean rainfall is 267 mm, yet the maximum recorded in one day is 145 mm.

Recorded world extreme values of meteorological elements have been published by various authors. An excellent coverage is provided by Riordan (1970), who also discussed the physical processes underlying the occurrence of each extreme type. A short summary of some of the most noteworthy events is included here to give some perspective to the overall question of extremes but the impact of such events is considered in greater detail in later chapters.

The world's lowest recorded temperature is $-88 \cdot 3$ °C at Vostok, Antarctica (78°27′ S, 106°52′ E, altitude, 3412 m) on 24 August 1960. Extreme low temperatures result from an optimum combination of several meteorological elements; absence of solar radiation, clear skies and calm air are the most essential requirements (McCormick, 1958). The ultimate fall in temperature is dependent on the duration of these conditions and is also affected by factors such as altitude and distance from the ocean.

The highest accepted recorded temperature is $58 \cdot 0$ °C at Al Aziziyah in Libya on 13 September 1922. Lamb (1958) lists the main influences for very high temperatures as being strong heating on dry desert sand or rock, a clear atmosphere, föhn effects, long stay

Table 2.7
WORLD OBSERVATIONS OF EXTREME RAINFALL
(after Paulhus, 1965; NZ Met. Service, 1976)

Duration	Depth (mm)	Location	Date
12 months	26 470	Cherrapunji, India	August 1860–July 1861
11 months	22 990	Cherrapunji, India	January–November 1861
6 months	22 454	Cherrapunji, India	April–September 1861
1 month	9 300	Cherrapunji, India	July 1861
15 days	4 798	Cherrapunji, India	24 June–8 July 1931
5 days	3 854	Cilaos, La Réunion	13–18 March 1952
2 days	2 500	Cilaos, La Réunion	15–17 March 1952
24 hours	1 870	Cilaos, La Réunion	15–16 March 1952
12 hours	1 340	Belouve, La Réunion	28–29 February 1964
9 hours	1 087	Belouve, La Réunion	28 February 1964
2 hours 45 minutes	558	D'Hanis, Texas	31 May 1935
42 minutes	305	Holt, Missouri	22 June 1947
8 minutes	126	Füssen, Bavaria	25 May 1920
1 minute	31	Unionville, Maryland	4 July 1956

or long passage of air over the warmest surface available and atmospheric subsidence to inhibit vertical convection and local circulations. The greatest temperature range over an unspecified period occurred at Verkhoyansk, USSR (67°50′ N, 133°50′ E), from −70 °C to 37 °C. The greatest temperature range in a day from 7 °C to −49 °C occurred at Browning, Montana, USA (48°35′ N, 113°10′ W) on 23–24 January 1916.

Some areas of the world are well known for their very heavy rainfalls and there are accepted meteorological reasons for these rains (Table 2.7). Persistent heavy rains throughout a season or year are generally associated with orographic lifting of persistent moisture-laden airstreams. Heavy rains for periods of days are usually associated with cyclonic storms, and short-period extreme falls with localized intense thunderstorms. Some of the extreme rainfall measurements for the world are listed in Table 2.7.

The maximum recorded surface wind gust of 108 m.s^{-1} (389 km.h^{-1}) was at Mt. Washington, New Hampshire, USA on 12 April 1934. The highest mean sea level pressure was 1084 mbar at Agata, Siberia, on 31 December 1968, and the lowest mean sea level pressure was 877 mbar, 965 km north-west of Guam on 14 September 1958.

SUGGESTIONS FOR FURTHER READING

BALDWIN, B., POLLACK, J. B., SUMMERS, A., TOON, O. B., SAGAN, C. and VAN CAMP, W., 1976, 'Stratospheric aerosols and climatic change.' *Nature*, 263, pp. 551–4.

BARNES, R. A., 1976, 'Long-term mean concentrations of atmospheric smoke and sulphur dioxide in country areas of England and Wales.' *Atmos. Env.*, 10, pp. 619–32.

BOLIN, B., 1974, 'Modelling the climate and its variations.' *Ambio*, 3, pp. 180–8.

BUDYKO, M. I., 1977, 'On present-day climatic changes.' *Tellus*, 29, pp. 193–204.

BUDYKO, M. I. and DAVITAJA, F. F., 1976, 'The influence of man on climate.' *Geoforum*, 7, pp. 99–106.

BURCHARD, J. K., 1975, 'Significance of particulate emissions.' *J. Air Polln. Control Assoc.*, 25, pp. 99–100.

BURROUGHS, W., 1978, 'On running means and meteorological cycles.' *Weather*, 33, pp. 101–109.

CLARKE, A. J. and SPURR, G., 1976, 'Routine sulphur dioxide surveys around large modern power stations. I – summary paper.' *Atmos. Env.*, 10, pp. 265–8.

CRUTZEN, P. J., 1974, 'Estimates of possible variations in total ozone due to natural causes and human activities.' *Ambio*, 3, 201–10.

DAVIES, C. N., 1974, 'Particles in the atmosphere—natural and man-made.' *Atmos. Env.*, 8, pp. 1069–80.

DEAGUE, T. K., 1975, 'Global atmospheric consequences of the combustion of fossil fuels.' *J. Inst. of Fuel*, 48, pp. 153–62.

FENNELLY, P. F., 1976, 'The origin and influence of airborne particulates.' *Amer. Scientist*, 64, pp. 46–56.

FLOHN, H., 1977, 'Climate and energy: a scenario to a 21st century problem.' *Climatic Change*, 1, pp. 5–20.

FRASER, P. J., 1978, 'The chlorofluoromethane (CFM) ozone controversy.' *Clean Air*, 12, pp. 6–11.

GALBALLY, I. E., 1976, 'Man-made carbon tetrachloride in the atmosphere.' *Science*, 193, pp. 573–6.

GERAKIS, P. A., 1974, 'Man-made disruptions of the carbon dioxide and oxygen concentration in the atmosphere and the role of plants.' *J. Environmental Quality*, 3, pp. 299–304.

GOLDSMITH, P., TUCK, A. F., FOOT, J. S., SIMMONS, E. L. and NEWSON, R. L., 1973, 'Nitrogen oxides, nuclear weapon testing, Concorde and stratospheric ozone.' *Nature*, 244, pp. 545–51.

HAMMOND, A.L., 1975, 'Ozone destruction: problem's scope grows, its urgency recedes.' *Science*, 187, 1181–3.

HAMPSON, J., 1974, 'Photochemical war on the atmosphere.' *Nature*, 250, pp. 189–91.

HARSHVARDHAN and CESS, R. D., 1976, 'Stratospheric aerosols: effect upon atmospheric temperature and global climate.' *Tellus*, 28, pp. 1–10.

HEINES, T. S. and PETERS, L. K., 1974, 'The effect of ground level absorption on the dispersion of pollutants in the atmosphere.' *Atmos. Env.*, 8, pp. 1143–53.

HERMAN, B. M. and BROWNING, S. R., 1975, 'The effect of aerosols on the earth-atmosphere albedo.' *J. Atmos. Sci.*, 32, pp. 1430–45.

HOBBS, P. V., HARRISON, H. and ROBINSON, E., 1974, 'Atmospheric effects of pollutants.' *Science*, 183, pp. 909–15.

HOFFERT, M. I., 1974, 'Global distributions of atmospheric carbon dioxide in the fossil-fuel era: a projection.' *Atmos. Env.*, 8, pp. 1225–49.

LANDSBERG, H. E., 1976, 'Concerning possible effects of air pollution on climate. Statement.' *Bull. Amer. Met. Soc.*, 57, pp. 213–15.

LANDSBERG, H. E. and ALBERT, J. M., 1975, 'Some aspects of global climatic fluctuations.' *Arch. f. Met. Geophys. u. Bioklim.*, B, 23, pp. 165–76.

LEWTHWAITE, G. R., 1966, 'Environmentalism and determinism: a search for clarification.' *Annals Assoc. Amer. Geogrs.*, 56, 1–23.

LORENZ, E. N., 1976, 'Nondeterministic theories of climatic change.' *Quaternary Res.*, 6, pp. 495–506.

MARTIN, A., 1974, 'The influence of a power station on climate—a study of local weather records.' *Atmos. Env.*, 8, pp. 419–24.

MITCHELL, J. M., 1976, 'An overview of climatic variability and its causal mechanisms.' *Quarternary Res.*, 6, pp. 481–94.

MORALES, C., 1977, 'Rainfall variability—natural phenomenon.' *Ambio*, 6, pp. 30–3.

NEWELL, R. E. and WEARE, B. C., 1976, 'Ocean temperatures and large scale atmospheric variations.' *Nature*, 262, pp. 40–1.

PARRY, M. L., 1975, 'Secular climatic change and marginal agriculture.' *Trans. Inst. Brit. Geogrs.*, 64, pp. 1–13.

PAULHUS, J. L. H., 1965, 'Indian Ocean and Taiwan rainfalls set new records.' *Mon. Weath. Rev.*, 93, pp. 331–5.

POLLACK, J. B., TOON, O. B., SUMMERS, A., Van CAMP, W. and BALDWIN, B., 1976, 'Estimates of the climatic impact of aerosols produced by space shuttles, SST's, and other high flying aircraft.' *J. Appl. Met.*, 15, pp. 247–58.

POTTER, G. L., ELLSAESSER, H. W., MacCRACKEN, M. C. and LUTHER, F. M., 1975, 'Possible climatic impact of tropical deforestation.' *Nature*, 258, pp. 697–8.

PRENTICE, S. A., 1978, 'Lightning risk.' *Search*, 9, pp. 222–9.

RAMANATHAN, V., 1975, 'Greenhouse effect due to chlorofluorocarbons: climatic implications.' *Science*, 190, pp. 50–1.

RATCLIFFE, R. A. S., WELLER, J. and COLLISON, P., 1978, 'Variability in the frequency of unusual weather over approximately the last century.' *Quart. J. Roy. Met. Soc.*, 104, pp. 243–56.

SCORER, R. S., 1976, 'A commentary on ozone depletion theories.' *Atmos. Env.*, 10, pp. 177–80.

TWOMEY, S., 1974, 'Pollution and the planetary albedo.' *Atmos. Env.*, 8, pp. 1251–6.

WATTS, R. G. and HRUBECKY, H. F., 1975, 'On the limits to energy growth.' *Technological Forecasting and Social Change*, 7, pp. 371–8.

WILLIAMS, J., KROMER, G. and GILCHRIST, A., 1977, *Further studies of the impact of waste heat release on simulated global climate: Part 1*. Research Memorandum, RM-77-15, International Institute for Applied Systems Analysis, Laxenburg, Austria).

WILLIAMS, J., KROMER, G. and GILCHRIST, A. 1977, *Further studies of the impact of waste heat release on simulated global climate: Part 2*. Research Memorandum, RM-77-34, International Institute for Applied Systems Analysis, Laxenburg, Austria).

WINSTANLEY, D., 1975, 'The impact of regional climatic fluctuations on man: some global implications.' In *Proceedings of WMO/IAMAP Symposium on Long-Term Climatic Fluctuations, Norwich, 18–23 August 1975*. WMO-No. 421, pp. 479–91.

3
Weather and Human Behaviour

3.1 HUMAN BIOMETEOROLOGY AND CLIMATIC DETERMINISM

The importance of the influences of weather and climate on human well-being and health has been well known since ancient times and many attempts have been made to establish clear and definite relationships, but the scientific study of human biometeorology is relatively new. The basic aim in human biometeorology is to try to assess how the human body reacts to changes in the atmospheric environment. An important part of human biometeorology is to establish how much of the overall biological variability is the result of changes in weather, climate, and seasons. Human biometeorological research, which is assuming growing importance and interest, is concerned with three broad areas of investigation: the influence on man of meteorological elements on physiological processes; the influence of weather and climate, as combinations of meteorological elements; and studies of requirements of man for modification of climatic stresses, through avenues such as clothing and buildings (Wallen, 1974). Components of the atmospheric environment likely to influence human behaviour and well-being include precipitation, temperature, humidity, solar radiation, wind, lightning, fog, clouds, atmospheric electricity, and both natural and man-made pollution.

Bates (1966) has distinguished three levels of climatic environment in relation to the role of weather in human behaviour. The 'microclimate' concerns conditions surrounding an individual organism; the ecological climate or 'ecoclimate' is that of the habitat, in the case of man, of the buildings in which he lives and works or of the fields that he cultivates; and the geographical climate or 'geoclimate' comprises the conditions measured by standard meteorological methods. Man has been modifying his microclimate for a long time by wearing clothing, and he has modified the ecoclimate even longer through the use of fire and the building of shelters. In addition he has long affected the geoclimate by such activities as clearing forests and cultivating land, building cities and polluting the atmosphere, but only recently has he thought of intentionally modifying climate.

Many decisions concerning human behaviour are clearly weather related. Some examples are listed in Table 3.1. Decisions related to weather criteria can be important and economically significant, but the real economic effect of the weather is difficult to assess. Many efforts have been made to determine optimal climatic conditions for human activities, but the results are not clear. Requirements vary from one activity to another and there is tremendous individual variation. Psychologists seem to have given little attention to the effects of weather and climate on behaviour. Certainly there are seasonal trends in

Table 3.1
SOME WEATHER-RELATED DECISIONS AND HUMAN BEHAVIOUR
(after Maunder, 1970)

Factor	Weather criteria	Decision	Adaptations
Recreation	Temperature Wind Precipitation	Desire Impulse	Clothing Rearrange time/place Different activity
Shopping	Various	Urgency of purchases	Transport Rearrange time/place
Travel (a) business	Visibility Precipitation Storms	Fastest means	Change time Change route
(b) pleasure	as above plus Cloud Wind Temperature	Where to go Route	Change time Change destination Change route
Voting	Temperature Precipitation	Civic duty Legal requirement Desire	Mobility Change time Do not vote
Work study	Various	Possible alternatives Degree of application Motivation	Protection Will-power Rearrange programme

many aspects of human behaviour, but it is difficult to determine whether such trends are related to weather conditions or to seasonal shifts in the economy. Weather conditions may produce reaction and medical conditions such as hay fever and asthma, or they may lead to a reduction in the labour force or a work slow-down, leading in turn to losses in productivity, earnings and taxes. Alternatively, costs may be incurred in modifying the environment of the worker or the patient to provide more satisfactory working and living conditions (Maunder, 1970).

Throughout his history man has been beset by the effects of atmospheric conditions, with slow fluctuations of climate causing migrations, extremes of seasonal weather bringing starvation, and various hazards leaving death and destruction. Landsberg (1971) made the point that in his natural state man can survive only in a very limited climatic zone. The present-day 25 °C annual isotherm encompasses about all the habitats in which our primitive ancestors could dwell, and the 1000 mm annual isohyet delineates areas with adequate year-round water supplies. Technology now exists to permit man's survival just about anywhere on earth; man is able to adapt himself through clothing and technology to many climatic circumstances which otherwise would be difficult to cope with.

The theory of climatic determinism, that there is a causal relation between stimulating climate and human inventiveness, is not widely supported today, but has been the subject of considerable debate during the twentieth century. The best-known advocate of the theory was Ellsworth Huntington (1945) who contended that a certain type of climate, now found mainly in Britain, France and neighbouring parts of Europe, and in the eastern USA, is favourable to a high level of civilization. Bates (1966) noted, however, that a most important factor is that civilization could not develop in higher latitudes until methods of coping with the cold and darkness of winter had been developed. From this it can be argued not that the climate of western Europe is stimulating, but that civilization was not possible there until methods had been found of overcoming climatic defects (Markham, 1947). In other words, the geoclimate remained the same, but the ecoclimate was greatly modified.

There is now little argument that weather and climate do influence human activities, not only in day-to-day terms but also over longer periods of climatic change. Perhaps there is a case for a reconsideration of some of the basic tenets of climatic determinism to redefine the concept, to allow for the increasing recognition given to the relationships between the atmospheric resource and man. There is an undoubted relationship between climate and the characteristics of any population (WHO, 1972). Through physiological and behavioural adjustments man is remarkably adaptable to his environment. Particularly severe climatic conditions may result in extensive adaptation, as in the case of people living at very high altitudes in the Andes or Tibet. Cyclic climatic changes influence biological rhythms, which enter into all human functions and activities. Man shows very wide individual variations in adaptability. Ethnic group differences appear as well as within-group variations.

In this chapter consideration is given to some of the ways in which interrelationships between weather and climate and human comfort, health and behaviour show themselves. This will permit a closer evaluation of the manner in which man might reach a more harmonious relationship with the atmospheric resource. If man is to make the best use of the resource he has to understand something of its direct effects upon himself. In subsequent chapters attention is turned to economic impacts, before looking at how man attempts to handle the resource at his disposal.

3.2 CLIMATE AND COMFORT

The main climatic variables determining physical discomfort are radiation, air temperature, humidity and wind. Complete assessment of physical discomfort also requires information such as thermal conductivity of clothing, vapour pressure of the skin, and the metabolic heat rate due to activity of the human body.

Both long and short wave radiation received by the body must be considered, together with the albedo of clothing and radiation emitted by the body and clothing. Clothing exerts an influence on the radiation balance through its colour and insulating properties. Exchange of sensible heat between the atmosphere and the body is influenced by both conduction and convection. Conduction is normally inefficient in adding heat to the body but it is an important factor in reducing heat loss by virtue of the low thermal conductivity of many clothing and building materials. Convection is an important cooling mechanism for the body, but its efficiency is restricted by conventional European-type clothing although enhanced by the freer-flowing garments of societies better adapted to warm conditions (Fig. 3.1).

Body temperature would rise by about 2 °C per hour if not offset by evaporative cooling from the lungs and skin, by convection if the air is cooler than the skin, and by radiation. In sustained work an adult may lose a litre of water in an hour, which would absorb energy

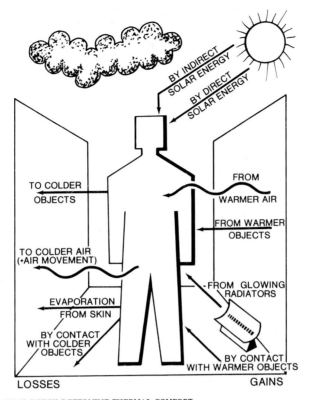

FIG. 3.1. HEAT INTERACTIONS WHICH DETERMINE THERMAL COMFORT
(After NCDC, 1977; reproduced with the permission of the Australian National Capital Development Commission).

and produce a cooling effect if evaporated. If evaporation is incomplete, there is less cooling and beads of sweat wet the skin. Wind is important because it accelerates heat transfer by turbulence and by evaporative cooling. It can prevent the accumulation of high moisture content next to the skin and thus help to maintain evaporative cooling close to the potential rate (Hounam, 1967). Wind can restore comfort to an overheated body, yet at the other end of the scale it can remove heat from a body and induce chill. It can also increase the heat load if the air temperature exceeds skin temperature.

Personal comfort depends on the maintenance of a body temperature of about 37 °C without excessive demands to increase heating by shivering or by exercise, or to increase cooling by evaporation (Linacre and Hobbs, 1977). Definition of discomfort criteria is difficult, however, because of human variations. Personal reactions to the weather differ greatly according to variables such as health, age, sex, clothing, occupation and acclimatization. This has not deterred many attempts to measure climatic strain, to derive discomfort indices based on the average response of subjects under specified conditions.

Any complete study of weather and physical comfort should ideally include several other meteorological factors. These include sequences of hot days, persistence of a weather type, and the possible relationships between weather and insects or air pollution.

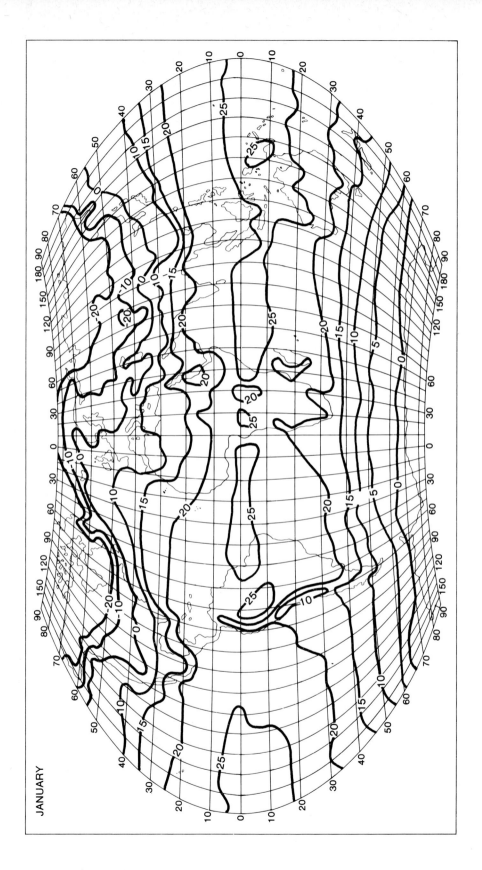

JANUARY

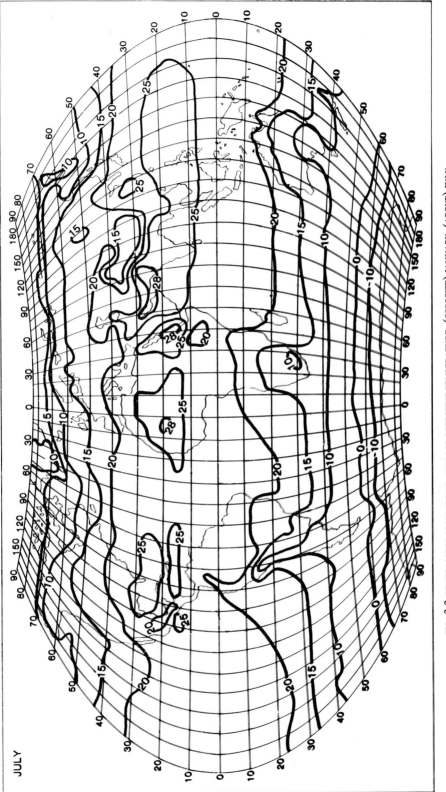

JULY

FIG. 3.2 GENERALIZED ISOLINES OF AVERAGE EFFECTIVE TEMPERATURE, (OPPOSITE) JANUARY, (ABOVE) JULY (After Gregorczuk and Cena, 1967; reproduced with the permission of Swets and Zeitlinger B.V.).

Insects such as mosquitoes affect comfort and have life cycles dependent on meteorological conditions, which also influence the development of uncomfortable levels of atmospheric pollution. In addition, in the case of high altitudes the atmosphere contains a lower density of oxygen. This means that the blood of a person at 3000 m holds 10 per cent less oxygen than at sea level and 30 per cent less at 4500 m. Apart from breathing difficulties, the deficit affects many functions of the body and the brain (Linacre and Hobbs, 1977). Factors like these can have considerable bearing upon human comfort, but are rarely, if ever, incorporated into commonly used comfort indices.

3.2.1 Comfort indices

Most physioclimatic indices are based on temperature, humidity or wind measurements, sometimes used singly, sometimes in combination. A widely used measure of comfort is known as the effective temperature (Fig. 3.2). This is defined as the temperature of still, saturated air which gives the same sensations of comfort to a normally clothed sedentary worker as are given by the actual conditions of temperature, humidity and wind.

Effective temperature is a useful and easily calculated index, usually derived from an appropriate nomogram. Its application is limited, however, because available criteria relate only to workers in sedentary occupations and because the significance of values more than 5 ° outside the comfort limits of 15–27 °C is not clear (Gaffney, 1973). Table 3.2 shows an effective temperature classification for Australia adapted from US experimental data. The lower comfort limit has been set at 15 °C rather than 17 °C to relate more closely to Australian conditions, where central heating is not general as in the USA. In actual environmental situations clothing variation helps to decrease discomfort on cold days, and workers also tend to move around more, thus increasing metabolic heat rates.

Table 3.2
AN EFFECTIVE TEMPERATURE COMFORT CLASSIFICATION,
ADAPTED FOR AUSTRALIA
(after Gaffney, 1973)

Effective temperature (°C)[a]	Comfort class
> 28·0	high discomfort
27·0–28·0	discomfort
25·0–26·9	transitional (warm)
17·0–24·9	comfort
15·0–16·9	transitional (cold)
< 15·0	discomfort

[a] indoor environment, normally clothed sedentary workers

Another measure of comfort is the heat stress index (Belding and Hatch, 1955), defined as the ratio between the amount of perspiration which must be evaporated from the skin to maintain thermal comfort and the maximum amount of evaporation which can occur under the particular conditions. A modified form of this relationship is the relative strain index, which takes account of the insulating effects of clothing and net radiation of heat to

Table 3.3
RELATIVE STRAIN INDEX COMFORT CLASSIFICATION
(after Gaffney, 1973)[a]

Relative Strain Index	Percentage of men comfortable/distressed
0.10	100 comfortable
0.20	85 comfortable
0.25	50 comfortable
0.30	0 comfortable
0.40	75 distressed
0.40	100 distressed

[a] for standard men (25 years, healthy, not acclimatized to heat) indoors, air movement 60 m.min^{-1}, metabolic heat rate generated by activity 100 kcal.m^{-2}.h^{-1} (420 kJ.m^{-2}.h^{-1})

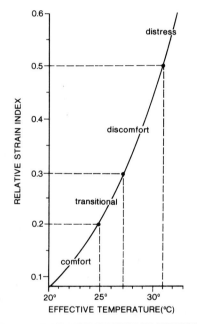

FIG. 3.3. RELATIONSHIP OF RELATIVE STRAIN INDEX WITH EFFECTIVE TEMPERATURE
(After Gaffney, 1973; reproduced with the permission of the Australian Bureau of Meteorology, Department of Science and the Environment).

the body. A discomfort scale based on this index is shown in Table 3.3. On this scale a relative strain index of 0·3 is taken as the critical point above which no person is comfortable (Fig. 3.3).

An interesting example of practical recognition of discomfort of inland and northern areas of Australia is the allowance of rebates on income tax according to zones of residence (Fig. 3.4). In 1976 the untaxed allowance in zone A was A$ 216 plus 25 per cent of other deductions for the maintenance of dependents, and in zone B A$ 36 plus 4 per cent.

The effects of weather on comfort are not restricted to the hotter and more humid areas of the world (e.g. articles by Hounam (1967), Jauregui and Soto (1967), Stephenson (1963), Tout (1977), and Watt (1967) on Alice Springs, Mexico, Singapore, London, and Bahrein respectively). Several optimum weather indices for temperate climates have been proposed, such as those for Manchester (Hughes, 1967), London (Poulter, 1962) and Armagh (Rackcliff, 1965). The concept of comfort is, however, subjective and measures such as the effective temperature and relative strain index are not as exact as they may appear, because they ignore many factors affecting the consciousness of comfort.

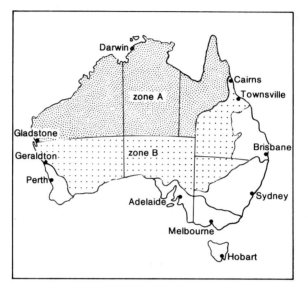

FIG. 3.4. ZONES OF AUSTRALIA WITH DIFFERENT INCOME TAX ALLOWANCES

Some of the recent research to develop a new and more powerful index of man's thermal discomfort is exemplified by the work of Gagge et al. (1972). They derived an index, called the standard effective temperature, based on simple physical theory and with a valid physiological basis (Gonzalez et al., 1974). Auliciems and de Freitas (1976) approached the problem of human comfort by considering clothing requirements for Canadians and thus attempting to describe climate in terms of human response. Their approach allows the application of energy balance principles and enables comparison of areas or seasons in quantitative terms related to man's behavioural response to climate.

Exposure to extremes of temperature and altitude may stress the human body beyond mere discomfort. For example, heat stroke occurs and death rates increase when body temperatures exceed 40 °C in effective temperatures above 33 °C (Linacre and Hobbs, 1977).

Windchill is a measure of the combined effects of low temperature and wind, related to excessive convective cooling of the bare skin. Frostbite occurs in Antarctica, for example, at −36 °C with a wind of 6 m.s⁻¹, but at −20 °C if the wind is 18 m.s⁻¹. When a clothed human is considered, account is taken of breathing and the insulation which the clothing provides for much of the skin surface. The thickness of clothing required to maintain thermal equilibrium provides a single index of the effect of cold and wind. Corresponding to any clothing thickness determined in this way there is a windchill equivalent temperature (Steadman, 1971). This is a measure of the cooling power of the wind, being the temperature of wind of 2·3 m.s^{-1} onto exposed flesh, producing the same sensation. The windchill equivalent temperature of calm air at 0 °C is +1 °C, for example, but it is −7 °C for a wind of 4·5 m.s^{-1} and an ambient temperature of 0 °C. For air at −10 °C, the windchill equivalent temperature in calm conditions would be −9 °C, and with a wind of 4·5 m.s^{-1} would be −19 °C.

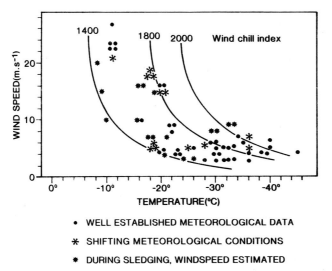

FIG. 3.5. FROSTBITE CASES RECORDED IN ANTARCTICA IN RELATION TO WINDCHILL BASED ON TEMPERATURE AND WINDSPEED
(After Wilson, 1967; reproduced with the permission of Swets and Zeitlinger B.V.).

It might be expected that windchill would be a problem in areas like Antarctica. Wilson's (1967) analysis of the correlation between windchill index and the occurrence of frostbite supports this (Fig. 3.5). What might be less expected is the occurrence of windchill in a country like Britain. In this regard Howe (1962) has presented an interesting analysis of the cold spell in December 1961 (Fig. 3.6), and Smithson and Baldwin (1978) have examined values of windchill for lowland sites in the UK, over a long period and for a single year.

3.2.2 Acclimatization

The human body can adapt to different climatic conditions by the process of acclimatization. This may take several weeks for young adults, and the ability to acclimatize seems to decrease with age. Some body adaptations, like the tanning mechanism of the skin, develop relatively quickly, others involve slow changes during growth (Bates, 1966). The physiologist is most concerned with the problems of human life in hot and cold climates and at high altitude, all essentially problems of the atmospheric environment.

Acclimatization to heat stress sets in very rapidly during the first days of exposure and is almost complete by the end of two weeks (WHO, 1972). The degree of acclimatization to heat may differ, depending on whether the environment is humid or dry, and may be better in the latter case because evaporation is then more effective. Acclimatization disorders seem to be more common in hot, humid environments. Studies have shown that when

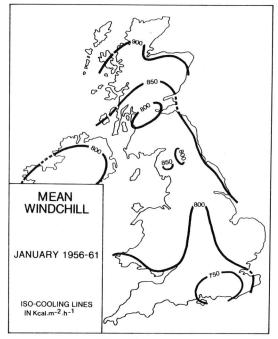

FIG. 3.6 MEAN WINDCHILL IN BRITAIN, JANUARY 1956–61
(After Howe, 1962; reproduced with the permission of the Royal Meteorological Society).

temperature rises alertness deteriorates and reaction, decision, and sensorimotor coordination times decrease (UNESCO, 1963; Lambert, 1968). These processes together account in part for the restrictive effect that climate may exercise on human activity in hot countries.

Sweating is the main thermoregulatory mechanism operating in hot climates. It serves to dissipate heat received from the environment as well as that produced by physical exertion. People from temperate zones who have never been exposed to tropical conditions can adapt physiologically to life in the heat, principally by alteration in their sweating mechanisms (Edholm, 1966). On first exposure to a hot and humid environment sweat rate is relatively low and body temperature progressively increases. Considerable discomfort and distress may lead to collapse. If exposure is continued day after day for only short periods each day, discomfort decreases, sweat rate increases and the rise of body temperature becomes smaller. This is the process of acclimatization to a hot environment. The important point is that complete acclimatization to heat can be achieved by relatively brief daily exposures, although the rest of the day may be spent in cool conditions. The critical factor seems to be that during heat exposure there must be sufficient stimulus for sweating.

Variations between ethnic groups appear to be small, in that an inhabitant of a temperate country acclimatizes as well as one of a tropical country.

Cold climates also stimulate thermoregulatory mechanisms, in this case to combat heat loss. The immediate response to cold is the initiation of a set of mechanisms to reduce heat loss and increase metabolic rate. Shivering, for example, is a particularly effective means of increasing heat production. A newcomer to a cold climate shows reduced shivering and metabolic heat production by the end of about ten days (WHO, 1972).

Many millions of people live and work at altitudes over 3000 m, where severe oxygen deficits and very low barometric pressure create many physiopathological problems. Sudden exposure to high altitude may provoke a variety of symptoms such as headache, insomnia, breathing difficulties and increased heart rate. Acclimatization is, however, fairly rapid and may be attained in about three weeks. People indigenous to high altitudes can make use of more oxygen in the air inspired than can acclimatized individuals, and their working capacity is comparable to that of men at sea level (Bouloux and Ruffié, 1971).

The nature and problems of acclimatization to temperature extremes and altitude have usually been examined in relatively severe conditions where changes are most likely to be apparent. There has to be an adequate stimulus to evoke physiological responses, and one of the problems is in determining what is an adequate stimulus (Edholm, 1966). There is still comparatively little known about reactions to relatively small environmental changes such as experienced from day-to-day or season-to-season. Even in a climate like that of Britain, the difference between summer and winter is sufficient to bring about a degree of acclimatization in warm weather. A very interesting consideration which seems to have received virtually no attention is that of acclimatization to different levels of atmospheric variability. Just as many climates are characterized by extremes of heat or cold, so also are many environments typified by varying amounts of diurnal, seasonal or annual variability requiring acclimatization for comfort.

3.2.3 Design for comfort

The provision of clothing, buildings, air conditioning, and heating are all ways of assisting adaptation to environmental conditions by creating artificial environments of greater comfort. They can be considered as forms of modification of the degree of acclimatization required, thus helping to alleviate acclimatization problems.

The main problem in building climatology is to what extent available climatological data, from meteorological stations not necessarily close to the building, may be applied for the purpose of design and construction in order to obtain an ideal indoor climate (Wallen, 1974). The problem basically involves the interrelation of the macroclimate derived from an ordinary meteorological station to the microclimate created in the immediate surroundings of a building. There has so far been little research into the nature of the microclimate within the 'sheath' around buildings.

Designing for comfort requires knowledge of appropriate criteria to specify conditions in which humans feel comfortable. Several indices of thermal comfort have already been discussed, including effective temperature. Studies have show that this overestimates the influence of humidity and that the radiant effect of surrounding surfaces is a more important factor when considering comfort conditions inside buildings. The environmental temperature is a measure which combines the effects of air temperature and the mean radiant temperature of surrounding surfaces. It is more useful than effective temperature when measuring thermal comfort in buildings (NCDC, 1977). In conditions with an absence of intense radiation sources and with random air motion, dry-bulb air temperature has been found to correlate as highly with thermal sensations as composite indices (Auliciems, 1977) (Table 3.4). This is useful since the measurement of mean radiant

Table 3.4
COMFORT LIMITS FOR SOME AUSTRALIAN LOCATIONS
(after Auliciems, 1977)

	Lower limit 80% comfort rate (°C)	Lower limit from regression (°C)	Upper limit from regression (°C)	Range from regression
Adelaide	19·5	18·5	23·0	4·5
Melbourne	20·5	18·9	22·0	3·1
Armidale	19·5	19·4	22·8	3·4
Perth	18·5	19·1	23·9	4·8
Brisbane	20·5	20·6	25·1	4·5

temperature presents some problems, although designers can usually take account of this effect by considering the types of building surfaces and the nature of insulation.

It is an established criterion that thermal comfort is not possible when the dry-bulb air temperature and the mean radiant temperature differ by more than 5 °C (NCDC, 1977). The effects of humidity and environmental temperature can be combined graphically to show comfort zones (Fig. 3.7). Comfort can be achieved with any combination of humidity and environmental temperature that falls within the zones defined for various activities in summer and winter. Clearly, there is a large amount of tolerance in the domestic situation where there is relative freedom to move about, change activity, or adjust clothing or heating/cooling amounts.

To maintain comfort conditions it is necessary to consume energy, especially in a cold climate. It is possible to reduce energy consumption by lowering comfort criteria, but this is perhaps an unrealistic approach. Alternatively, conservation measures may be adopted to minimize heat loss in winter and heat gain in summer, to maximize any benefits that might be derived directly from the environment (radiant solar energy), and to optimize the winter heating needed to satisfy the differential between desired comfort level and heating provided through other measures.

The main design principles for temperate climates common to both summer and winter are the use of insulation and thermal mass. Insulation provides for prevention of both excessive heat loss in winter and heat gain in summer, as well as raising the mean radiant temperature during winter. Thermal mass can provide a heat storage for solar radiation in winter, and if shaded in summer it can act as a heat sink to lower room temperature. Concrete slabs, for example, act as very efficient thermal masses. Ventilation should be controlled to minimize draughts in winter but to allow cross-ventilation in summer. Differing summer and winter requirements produce conflicts in design principles for shape, orientation, eaves and overhangs of buildings, and for landscaping of their surrounds. A cube would be the most efficient shape in terms of minimal heat loss, but the use of passive solar energy in winter and the avoidance of western insolation in summer

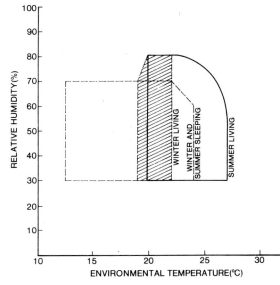

FIG. 3.7. COMFORT ZONES FOR DWELLINGS
(After NCDC, 1977; reproduced with the permission of the Australian National Capital Development Commission).

indicates an optimum shape with a ratio of 1·5 to 1·6 : 1·0 for north sides : east sides (NCDC, 1977) (Fig. 3.8).

In practice most houses are not designed with low energy requirements in mind, thus leading to the relatively wasteful use of various forms of heating and air conditioning to promote several objectives, including cooling, ventilation, air purification, dehumidification, and circulation. The air conditioner can thus be very valuable in increasing the efficiency of people in domestic or work situations during hot and humid weather. Air conditioning is vital in many key areas, such as hospitals, computer complexes, various industrial plants and offices. It is particularly valuable for patients suffering from respiratory tract ailments and is important for the efficient operation of restaurants, theatres and stores. In these situations the presence of air conditioning may be a positive factor in attracting custom away from non-air conditioned, hence less comfortable, premises.

This brief consideration of some of the factors involved in designing for comfort has indirectly pointed to the significance of weather and climate parameters for architects, engineers, builders and home owners. In addition there are obvious economic implications for the power supply industry, builders' suppliers, and manufacturers, wholesalers, retailers, and purchasers of air conditioning and heating equipment.

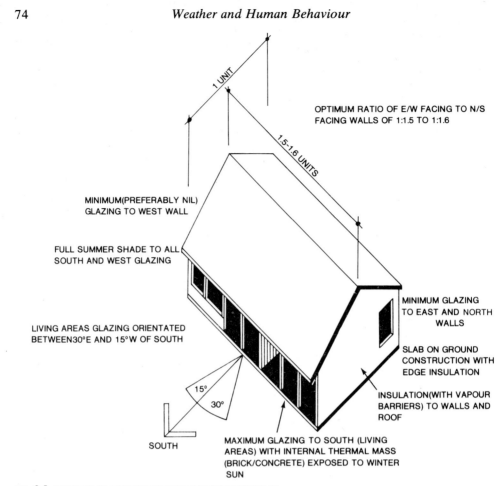

OPTIMUM RATIO OF E/W FACING TO N/S
FACING WALLS OF 1:1.5 TO 1:1.6

1 UNIT

1.5-1.6 UNITS

MINIMUM(PREFERABLY NIL)
GLAZING TO WEST WALL

FULL SUMMER SHADE TO ALL
SOUTH AND WEST GLAZING

MINIMUM GLAZING
TO EAST AND NORTH
WALLS

LIVING AREAS GLAZING ORIENTATED
BETWEEN30°E AND 15°W OF SOUTH

SLAB ON GROUND
CONSTRUCTION WITH
EDGE INSULATION

15°

30°

INSULATION(WITH VAPOUR
BARRIERS) TO WALLS AND
ROOF

SOUTH

MAXIMUM GLAZING TO SOUTH (LIVING
AREAS) WITH INTERNAL THERMAL MASS
(BRICK/CONCRETE) EXPOSED TO WINTER
SUN

FIG. 3.8. SUMMARY OF LOW ENERGY HOUSE DESIGN PRINCIPLES
(After NCDC, 1977; reproduced with the permission of the Australian National Capital Development Commission). This figure illustrates the situation in the Northern Hemisphere. In the Southern Hemisphere maximum glazing should face north.

3.3 CLIMATE AND HEALTH

It is a logical progression from considerations of comfort to those of health, since discomfort may aggravate illness or make the difference between sickness and health. Although the field of medical climatology is large and complex, the effects of climate on disease are not well understood. Comprehensive reviews such as those by Sargent and Tromp (1964) and Tromp (1963) point to the possible harmful effects of certain climatic factors on respiratory conditions, rheumatic diseases, skin cancer and cardiovascular disorders. In addition researchers have often described the possible role of winds like the föhn, sirocco, and sharav as causes of irritability, depression, headache, dizziness, haemorrhage, and hypertension (WHO, 1972; Winstanley, 1972).

3.3.1 Diseases and illnesses

Respiratory ailments are the most common form of illness and also show some of the closest links with atmospheric conditions. Asthma, in particular has received considerable attention. Derrick (1965, 1966) demonstrated the relation between seasonal and annual variations of asthma cases in Brisbane. Monthly attendances for hospital treatment from May to October, when the mean temperature is normally less than 21·1 °C, showed a significant correlation with the temperature one to two months earlier. For weekly attendances there was a highly significant correlation in each year with the mean temperature six weeks earlier. A more detailed examination of weekly attendance in relation to weather variables showed that high-asthma weeks tended to be associated with decreased mean and minimum temperatures, dew point, relative humidity and rainfall, but with an increase in sunshine hours (Derrick, 1969). There was a significant association with the arrival of a cold, dry change. No significant difference was found to be associated with smoke density, fungal elements or pollen. The increase in illness generally occurred within 48 hours of a definite fall in minimum temperature and similarly for humidity. However, some weeks of high asthma incidence did not follow a cold or dry change, and many cold, dry changes were not followed by an increase in asthma. The low-asthma weeks on average had higher dew points and relative humidities, more rainy days, lower temperature ranges and fewer hours of sunshine.

Evidence from other parts of the world seems to support exposure to cold as the most common weather-related factor provoking asthma attacks. Greenburg et al. (1964, 1967), for example, observed a statistically significant increase in asthma cases at three New York hospitals with the first or second cold periods in autumn. Tromp and Bouma (1965) reported that the number of asthma attacks in children, seven to sixteen years old in the eastern Netherlands, increased during periods of cooling. Spodnik et al. (1966) found that the airflow resistance of 100 Baltimore students increased significantly as the temperature decreased. On balance there seems little doubt that some asthmatic attacks may be directly provoked by cold, but it is uncertain whether this is related to inhalation of cold air or the exposure of the body to winter cold. Increased asthma might also be a direct result of irritation by dry air, but is difficult to separate the effects of dry air and cold air. A cold change might produce an asthmatic attack by first inducing a respiratory infection.

The occurrence of asthma and other respiratory diseases may also be dependent upon the abundance of allergens, variations of which may follow certain weather changes. The release of pollen into the air is triggered by a fall in relative humidity, while conversely, high humidity may lead to the accumulation of moisture on air-borne particles and accelerate their deposition (Stern, 1968). Paulus and Smith (1967) showed that attacks of bronchial asthma among Minneapolis students periodically exposed to air-borne pollutants from grain processing and storage installations were twice as frequent in students who were grain-sensitive or allergic to the dust.

Many other associations between weather and the incidence of various ailments are readily apparent although not so easy to explain. Rapid changes of atmospheric pressure have been found to correlate with the incidence of perforated duodenal ulcers (Hansen and Pedersen, 1972) and peripheral arterial embolism (Hansen, 1970). Davis (1958) demonstrated a close agreement between marked variations in temperature and the frequency of haemorrhaging from duodenal ulcers. A warm climate with relatively little daily and seasonal change in temperature would appear to afford the most suitable residence for those prone to suffer from duodenal ulcers. Davis made the point that there is a case for sufferers of duodenal ulcers to take notice of forecasts of sudden falls in temperature.

In the Netherlands and in Australia there is a tendency for people who later develop cancer of various kinds to have been born in winter (De Sauvage-Nolting, 1968). This may be because of the quality of the winter sunlight. Insufficient sunlight prevents the formation of adequate vitamin D in the body, necessary for the assimilation of calcium and phosphorous needed for bone growth (Linacre and Hobbs, 1977). There is a similar high incidence of people born in winter among those subsequently developing mental deficiency or schizophrenia, but the reason for the correlations is hard to determine.

Many diseases have a seasonal occurrence, which is probably related to climatic conditions. For example, scarlet fever, diphtheria and jaundice occur in Switzerland mainly in winter, whereas measles, influenza and chicken pox are most common in spring. Respiratory and heart complaints are at a maximum in late winter and early spring in England and Australia. Some infectious diseases like bubonic plague are highly dependent on the weather. Olson (1969, 1970) reported that the incidence of bubonic plague in Vietnam is inversely proportional to rainfall. The vector transmitting the disease is a flea carried by rats. The fleas become more numerous in dry weather when the monthly rainfall is less than about 100 mm, followed by a resurgence of the plague about two months later.

The transmission of other diseases by insects is also influenced by weather conditions. Mosquitoes, for example, bite and feed during rising temperatures or soon after, and the stability of the virus they transmit itself depends on temperature. The severity of an outbreak of an insect-borne disease depends mainly on the abundance and density of the insect transmitting the disease. These in turn are affected by a combination of climatic factors during survival and breeding stages and by variations in weather that help to concentrate the insect population in favourable habitats. The mosquitoes which transmit malaria exist only where temperatures are always above 15 °C and where the rainfall is over 1000 mm.a^{-1}, to provide the stagnant water needed for breeding. The organism causing yellow fever is also carried by a mosquito and is unable to develop at temperatures below 20 °C (Linacre and Hobbs, 1977).

Health, disease and economic activity are closely associated. Various associations between weather and health and disease are also apparent, so it is reasonable to suggest that variations in the atmospheric resource such as those produced by man-made pollution, through their effects on health and disease must have a direct bearing on economic activity (Maunder, 1970). Many studies from around the world have shown a close association between air pollution and ill health, particularly for diseases like bronchitis and lung cancer and less so for cardiovascular diseases and non-respiratory tract cancers. Lave and Seskin (1970) estimated the annual cost of respiratory disease in the USA as approaching $ 5000 million and emphasized the rising costs of increased sickness and mortality due to air pollution.

The physiological consequences of a pollutant depend on the dose, which is proportional to the concentration and the exposure. The first noticeable effect of photochemical pollution is a stinging of the eyes due to an acrid gas called peroxyacetylnitrate, formed from reactions between hydrocarbons, nitrogen oxides and atmospheric oxygen under the influence of sunlight. Ozone, which is also formed in these reactions, damages lung tissue and increases death rates as a result of swellings in lung passages. Athletic performance is reduced in ozone concentrations of only 0·13 mg.m^{-3}. Sulphur dioxide can lead to infections of the lower respiratory tract, especially among the elderly, the very young and those already weakened by illness. The effect of sulphur dioxide pollution is much worse if accompanied by dust or smoke (Fig. 3.9). The effects on health of some of the most common air pollutants are summarized in Table 3.5.

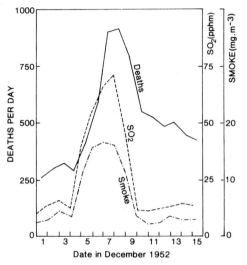

FIG. 3.9. EFFECTS OF POLLUTION LEVELS ON DEATHS IN LONDON, DECEMBER 1952
(After Royal College of Physicians, 1970; reproduced with the permission of Pitman Medical Publishing Co.).

Table 3.5
SOME EFFECTS OF POLLUTANTS
(after Linacre and Hobbs, 1977)

Pollutant	Approximate concentration (mg.m^{-3})	Duration of exposure	Effects
Sulphur dioxide	2·86	—	smell detectable
	0·43	1 week	certain plants visibly damaged
Carbon monoxide	438	1 hour	severe headache, nausea, collapse
	125	1–2 hours	reduced vigilance
Nitrogen dioxide	10·25	2 months	susceptibility to pneumonia (in monkeys)
	2·05	1 hour	lung proteins affected (in rabbits)
	0·41	—	smell detectable
Ozone	2·14	½ hour	headache, drastic reduction of oxygen transfer to blood
	1·07	3 months (3 hours per day)	obstructions in lungs, chest constriction, throat irritation, tight facial skin, lethargy
	0·21	—	dry nose membranes
	0·17	—	susceptibility to streptococcus bacteria

3.3.2 Mortality

It is clear from Figure 3.9 that excessive amounts of atmospheric pollution seem to increase the death rate, and there are several frequently quoted examples from around the world to support this (e.g. McDermott, 1961; Royal College of Physicians, 1970; Padmanabhamurty, 1972; Lave and Seskin, 1973a). There have been many studies of the relationships between human mortality and other aspects of the atmosphere. In particular, attention has been given to possible associations between mortality and seasonal weather conditions, short-term periods of atmospheric turbulence and severe heat stress.

Tromp (1963) summarized much of the work to 1963 in terms that are still relevant. He concluded that the mortality rate is higher in winter than in summer, at least for parts of Europe and the USA; that mortality increases during periods of great atmospheric turbulence, with the sudden influx of cold polar air and with the passage of active weather fronts; and that mortality from respiratory diseases increases rapidly during foggy weather when temperatures fall. Rosenwaike (1966) showed that death rates of both influenza and pneumonia sufferers were twice as high in January as in August in the USA. Similar, but less marked patterns were found for cardiovascular and renal diseases. Excessive mortality in winter has also been reported from Britain, Sweden and Australia, although the seasonal fluctuations are less marked in areas like Hawaii and California, where seasonal temperature differences are relatively small (Goldsmith and Perkins, 1967). There is also evidence to suggest, however, that seasonal mortality peaks may be disappearing because of improvements in public health and preventive medicine (Momiyama, 1968).

The importance of short-term weather variations on mortality has been clearly demonstrated by Driscoll (1971). He examined about 100 previous studies and found that 83 per cent of them showed some positive relationship between mortality and short-term weather variations, mostly related to the effects of changing and turbulent weather. Driscoll studied daily weather-related deaths in ten large cities in the USA and found that although total mortality correlated best with weather events, deaths of those over 70 also correlated well (Fig. 3.10). It is interesting to note that a case study of a cold front passage across the USA in the autumn of 1963 showed no apparent increase in mortality, so the relationship between mortality and disturbed atmospheric conditions does not necessarily always apply (Driscoll and Landsberg, 1967).

Severe stress and mortality have long been linked together, but it is difficult to find objective measures of heat stress. In general, heat-related deaths occur most frequently among the aged and increase with temperature above a vague critical value. Exposure time is an important factor. Most excess deaths do not occur until at least 24 hours after the onset of extreme heat. Another important factor in heat-related ailments is the temperature to which the body is customarily exposed or acclimatized. For example, a mean temperature of 32 °C over several days normally has little effect on healthy individuals used to such temperatures, whereas it may be fatal to those accustomed to only 27 °C. Severe heat stress will usually provoke fatalities, particularly in late spring or early summer when a sudden rise in temperature may occur before the body has had an opportunity to adjust its thermoregulatory system.

There are several examples from around the world to demonstrate the apparent rela-

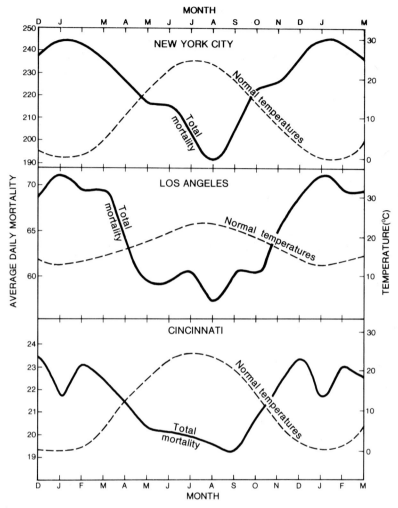

FIG. 3.10. ANNUAL MORTALITY CURVES FOR THREE CITIES IN THE USA
(After National Center for Air Pollution Control, 1967).

tionships between urban mortality and heat waves. A temporary increase in mortality in
Greater London, for example coincided with a heat wave in late July and early August
1975, and again during late June and early July 1976 (Macfarlane and Waller, 1976; Tout,
1978) (Fig. 3.11). Increases in mortality associated with heat waves have been widely
reported in North America. There were over 500 excess deaths in St Louis in July 1966
and over 100 in Illinois and New York City at the same time (Bridger and Helfand, 1968).

The death rates from heat-related ailments are usually much higher in urban areas than
in the country. This may be the result of climate modification due to urbanization (e.g.
Clarke, 1972; Padmanabhamurty, 1972). Daytime urban-rural differences are small, but
night-time differences are often significant: cities are warmer, long wave radiant heat load

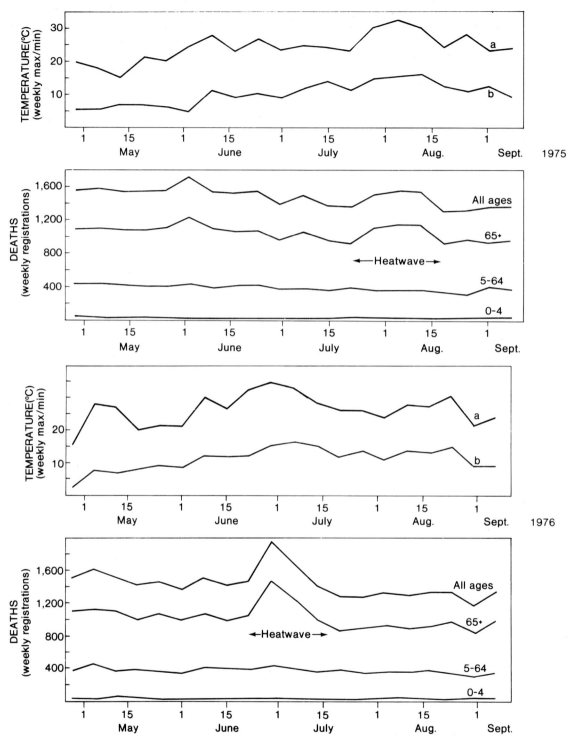

FIG. 3.11 WEEKLY TOTALS OF DEATHS REGISTERED IN GREATER LONDON BY AGE AND WEEKLY MAXIMUM (a) AND MINIMUM (b) TEMPERATURES (LONDON WEATHER CENTRE) FOR SUMMER 1975 AND 1976
(After Macfarlane and Waller, 1976; reproduced with permission of Macmillan Journals Ltd. and Dr A. Macfarlane).

is greater, windspeeds are often lower, and inside temperatures are often higher. Nocturnal urban heat islands of 4 to 7 °C are known to be associated with heat waves. During heat waves urban dwellers may experience sustained thermal stresses, day and night, whereas the rural inhabitant probably only experiences them during the day.

3.4 WEATHER AND SOCIETY

It has been demonstrated that many human physiological reactions are closely associated with the state of the atmosphere. Increased awareness of matters like air pollution problems and the growth of research into atmospheric hazards have been accompanied by a growing interest in sociological reactions. Haas (1968) reviewed the sociological aspects of human reactions to the atmosphere and suggested several major areas of research interest, including the incidence of crime, the nature of recreational and leisure time activities, the disruption and use of basic community services, the demand for and use of welfare services, and participation in political activity.

The frequency with which the weather provides the topic for opening a conversation or for superficial comment emphasizes the ways in which it affects our daily lives in terms of things like choice of clothing, method of transport and allocation of time to various activities. Yet this very familiarity tends to mean that most people ignore the importance of a thorough investigation of the influence of weather on human activity patterns.

There is some evidence that increased mental stress and the incidence of crimes such as homicide and assault may be associated with certain types of weather conditions. Warm, dry and very strong winds such as the föhn and chinook of mountain areas or the sirocco and sharav of the Mediterranean region may increase mental stress, with possible sociological repercussions (Winstanley, 1972). In particular, Miller (1968) has related the daily homicide rate in the Los Angeles area to the incidence of the hot, dry Santa Ana wind.

The US Federal Bureau of Investigation noted what seemed to be a direct and significant relationship between seasons and the incidence of some types of crime. Climate was one of eleven conditions listed by the Bureau as affecting the amount and type of crime. The so-called long hot summer, particularly in some of the largest cities of the USA, seems to coincide with national annual peaks in various crime categories. There seems to be scope for considerably more work on weather-related aspects of crime statistics, since this could assist law enforcement authorities to make the most effective use of available resources (see Maunder, 1970, and Miller, 1968, for more detailed discussions of weather and crime). Maunder (1970) has also indicated an apparent relationship between weather and rioting, especially race riots in the USA, although there could be a broader connection between civil disorder and the weather. He noted that while heat and humidity appear to be associated with rioting, it is also quite clear that atmospheric cooling may help to stop the disturbance. Rain also seems to remove the tensions incurred from monotonous hot and humid weather, and has been reported as a factor in ending rioting.

It might be expected that stress-producing weather situations would also be influential on the suicide or attempted suicide rates (Tromp and Bouma, 1972). In fact there is little evidence of such a connection, but Rosenwaike (1966) noted that the seasonal curve of deaths from suicide showed a marked peak in spring and a trough towards the end of the year. This suggests an association with turbulent weather rather than monotonous and oppressive conditions.

Recent investigations by William Lyster, a research biologist, have unearthed a possible relationship with sociological implications between the sex ratio for births in Britain and major weather events. He found a brief boom in boy babies in August 1977 in several parts of Britain, in some cases giving ratios of 140 boys to 100 girls, compared with the normal 106 boys to 100 girls. He has attributed this marked variation to the after-effects of the British drought of 1975–76. The hypothesis is that when a drought breaks there is a surge of accumulated trace elements into the water supply, with unusually high concentrations affecting the formation of spermatozoa. Spermatozoa usually take about 50 days to mature, which added to the normal 9 month gestation period for a human baby leads to a 320 day delay between the effect and the cause. If the starting date in the case of the 1975–76 drought is taken as mid-September, two weeks after the rains started, to allow for replenishment of huge moisture deficits before large-scale runoff developed, a boy baby boom should have occurred in August 1977, as reported. Lyster also found another possible connection between a major weather episode and unusual birth sex ratios in some London hospitals. The great London smog of December 1952 was followed about 320 days later by unusual birth ratios of 109 boys to 144 girls. There have also been suggestions that polluted air might be responsible for the predominance of girls among the children of male anaesthetists and workers in PVC plants and foundries (*Sunday Times*, London, 9 July 1978).

Many aspects of modern living are geared to weather conditions (Maunder, 1970). Man's inadvertent or deliberate modifications of the atmosphere raise important questions of adjustment and readjustment with economic, social, legal, and political implications (chapter 8). The perception of weather extremes and hazards and the adjustments involved also touches on basic sociological considerations (chapter 4).

In 1976 the American Meteorological Society published a set of statements about the atmospheric sciences and problems of society, as a contribution to the continuing assessment of the social impact of meteorology in the years ahead. Several of the main points made are worth summarizing here, since they help to emphasize the significance of relationships between weather and human activities, such as those discussed in the following chapters.

According to Dale (1976) weather is the most important uncontrolled variable affecting food production today. Agriculture and forestry deal with renewable resources which are largely generated and controlled by our climatic resources. Meteorological information is essential for the management decisions that must be made for intelligent planning at all levels, from soil and crop management decisions for a single farm to estimation of the proportion of national grain production available for export. The geneticist looking for better and more adapted crop varieties, the designer of farm equipment, the manufacturers of fertilizers and the whole chain of salesmen and transporters of agricultural inputs vital to high levels of management must look to the meteorologist for information with which to make their decisions.

Mechanization, crop hybrids, chemical fertilizers, pesticides and irrigation have combined to influence many people to consider that crop production in countries like the USA is independent of the weather. In fact, as the USA summer drought of 1974 demonstrated, technology has not removed the weather variable from crop production. Indeed, the implication of weather-related decreases from expected crop yields is far more serious when yield expectations are high because of the application of technology.

Golden and Abbey (1976) emphasized the relevance of severe local storms to societal problems at the national and international levels and pointed to the need for improved understanding of severe storm behaviour and characteristics. Crawford (1976) similarly pointed to the relevance of improved understanding and predictability of atmospheric turbulence, particularly in relation to effects on aircraft and their passengers and the patterns of turbulence involved in design of buildings and urban planning. The dependence of around 300 million United States airline passengers in 1975 on the meteorological profession for safe and efficient completion of flights was stressed by Bromley (1976).

Franceschini (1976) acknowledged inflation and unemployment as the pressing problems of society, but suggested that the major societal problems of national and international scope, concerning the general welfare of mankind in the present and future, are directly associated with energy generation, food supply, environmental pollution and the prediction of weather and climate. The last has assumed new proportions as a consequence of the others.

Bosart (1976) stressed that weather recognizes no international boundaries and is closely related to every aspect of human behaviour, and that the health, safety and well-being of people are dependent upon the daily, weekly, monthly, seasonal, and yearly vagaries in weather patterns. Weather forecasts are required for everyday decisions in most economic pursuits, including agriculture, industry, and commerce. Short-term forecasts of potentially destructive weather phenomena are necessary to ensure the safety of people and property. Large departures from long-term climatological patterns can produce considerable economic and personal hardship. Hence weather forecasting is important and relevant to societal problems, locally, nationally, and internationally. Important sociological questions relate to the field of weather forecasting in terms of dissemination, use, interpretation and understanding of forecasts (chapter 7). The possible impact of long-term climatic changes on human societies is of major importance and will require international cooperation (chapter 8).

Some of the relationships and ramifications of the associations between weather and society are examined in the following chapters, which look at the ways in which the atmosphere influences man's activities, together with the adjustments he makes. The final two chapters consider how man handles the resource at his disposal in efforts to achieve greater harmony, economy and comfort.

SUGGESTIONS FOR FURTHER READING

AULICIEMS, A., 1976, 'Weather perception—a subtropical winter study.' *Weather*, 31, pp. 312–16.

—1969, 'Effects of weather on indoor thermal comfort.' *Internal. J. Biomet.*, 13, pp. 147–62.

BITAN, A., 1976, 'Wind as a negative factor in human comfort and its implications for planning.' *Internl. J. Biomet.*, 20, pp. 174–83.

COLE, R. J., 1977, 'Climate and building design.' *Weather*, 32, pp. 400–6.

COURT, A., 1948, 'Windchill.' *Bull. Amer. Met. Soc.*, 29, pp. 487–93.

FINKELSTEIN, J., 1971, 'Climate and comfort in the tropical South Pacific islands.' *N.Z. Geogr.*, 27, pp. 56–64.

FREESTON, D. H., 1974, 'Wind environment of buildings.' *Town Planning Quarterly*, 35, pp. 33–7.

FRISANCHO, A. R., 1975, 'Functional adaptation to high altitude hypoxia.' *Science*, 187, pp. 313–19.

GENTILLI, J., 1978, 'Physioclimatology of Western Australia.' *Geowest*, 12, (Department of Geography, University of Western Australia).

HOPKINS, J. S. and WHYTE, K. W., 1975, 'Extreme temperatures over the United Kingdom for design puposes.' *Met. Mag.*, 104, pp. 95–102.

LANDSBERG, H. E., 1969, *Weather and health*, (Doubleday).

—1973, *The assessment of human bioclimate, a limited review of physical parameters.* Technical Note No. 123, WMO-No. 331 (World Meteorological Organization, Geneva).

MACFARLANE, W. V., 1958, 'Thermal comfort zones.' *Architectural Science Rev.*, 1, pp. 1–14.

McINTYRE, D. A., 1976, 'Thermal sensation. A comparison of rating scales and cross modality matching.' *Internl. J. Biomet.*, 20, pp. 295–303.

McLAUGHLIN, J. T. and SHULMAN, M. D., 1977, 'An anthropocentric summer severity index.' *Internl. J. Biomet.*, 21, pp. 16–28.

MENZIES, J. B., 1971, 'Wind damage to buildings.' *Building*, 221, pp. 67–76.

MORGAN, D. L. and BASKETT, L. R., 1974, 'Comfort of man in the city. An energy balance model of man–environment coupling.' *Internl. J. Biomet.*, 18, pp. 184–98.

OLGYAY, V., 1967, 'Bioclimatic orientation method for buildings.' *Internal. J. Biomet.*, 11, pp. 163–74.

PIELKE, R. A., 1975, 'Influence of the sea breeze on weather and man.' *Weather*, 30, pp. 208–21.

PRESTON-WHYTE, R. A., 1975, 'A note on some bioclimatic consequences of coastal lows.' *S. Afr. Geogrl. J.*, 57, pp. 17–25.

SAINI, B. S., 1973, *Building environment. An illustrated analysis of problems in hot dry lands*, (Angus and Robertson, Sydney).

SHELLARD, H. C., 1967, 'Wind records and their application to structural design.' *Met. Mag.*, 96, pp. 235–43.

SIBBONS, J. L. H., 1966, 'Assessment of thermal stress from energy balance considerations.' *J. Appl. Physiol.*, 21, pp. 1207–17.

SMITH, V. K., 1976, 'The measurement of mortality—air pollution relationships.' *Environment and Planning*, A, 8, pp. 149–62.

SZOKOLAY, S. V., 1976, 'Thermal controls in northern Australia.' *Architectural Science Rev.*, 19, pp. 58–60.

TERJUNG, W. H., 1967, 'The geographical application of some selected physio-climatic indices to Africa.' *Internl. J. Biomet.*, 11, pp. 5–19.

—1967, 'Annual physioclimatic stresses and regimes in the United States.' *Geogrl. Rev.*, 57, pp. 225–40.

—1966, 'Physiologic climates of the conterminous United States: a bioclimatic classification based on man.' *Annals Assoc. Amer. Geogrs.*, 56, pp. 141–79.

TURNER, A. E., 1978, 'Discomfort in Bahrain.' *Weather*, 33, pp. 334–8.

WALSH, P. J., 1976, 'Energy requirements of Australian dwellings for heating and cooling.' *Aust. Refrigeration, Air Conditioning and Heating*, 30, pp. 9–13.

4
Impact of Extreme Events

Reference was made in chapter 2 to the importance of atmospheric hazards and extreme events. In the present chapter this theme is developed in greater detail, particularly in relation to impacts on human activity. Clearly, it would be impossible to examine the whole range of hazards and extreme events, so examples have been selected to illustrate the nature and dimensions of the problems posed.

Three main facets are considered in the following sections: the impact of specific phenomena such as tropical cyclones, tornadoes and droughts; the possible effects of atmospheric hazards and extremes on a particular form of economic activity, using transport as the example; and the question of climatic change and world food production, since changing climates are being viewed increasingly as a major threat to agricultural productivity being able to meet growing demands.

4.1 RANGE OF EVENTS

It is difficult to distinguish between atmospheric and non-atmospheric factors producing natural hazards. For example, hot, dry winds promote fire disasters but do not cause them. Similarly, an avalanche depends on the quality and quantity of snow and on the timing of a thaw, but it is unlikely to happen without certain slope characteristics.

Atmospheric phenomena may be responsible for coastal erosion, flooding, physical damage to property, disruption of transport and general economic and social activity, destruction of crops and animals, initiation of bushfires and forest fires, and the taking of human lives. The range of atmospheric hazard effects is much greater than is generally appreciated. Reports from around the world include, for example, lightning welding solid an iron joint in the leg of a cricket umpire; lightning tearing off part of the wing of a Boeing 747 approaching Tokyo; a sudden thaw in Chicago causing huge chunks of ice weighing up to 9 kg to fall 300 m or more from tall buildings; giant hailstones being responsible for the crash of a DC-9 in Georgia. Economic impacts of even relatively small-scale, isolated events can be large, as the Australian examples in Table 4.1 show.

In themselves weather extremes are not necessarily hazardous, although they may become so if they prevail for prolonged periods of time. This accumulative effect of weather extremes is evident in the cases of droughts, heatwaves and floods. Atmospheric factors may fulfil a variety of roles in the development of a hazard situation. A broad distinction can be made between phenomena such as tropical cyclones, tornadoes and

Table 4.1
SOME EXAMPLES OF RECENT ATMOSPHERIC HAZARDS IN AUSTRALIA
(based mainly on information contained in Annual Reports of the Bureau of Meteorology)

Feature	Year	Location	Effects (values are estimates)
Drought	1969–70	Widespread	25% slump in net farm income
TC Glynis		Dampier, W.A.	Breakwater damage
TC Ingrid		W.A.	Flooding, losses $1 million
TC Ada		Queensland Central Coast	13 deaths, damage $12 million
Tornado		Bulahdelah, N.S.W.	1·5 million trees destroyed or damaged
Lightning	1970–71	N.S.W.	3 deaths
TC Eva		Broome, W.A.	Extensive damage
TC Dora		Queensland South Coast	Damage $1 million
TC Sheila		Roebourne, Dampier W.A.	220 km.h^{-1} wind gusts
Strong winds		Albany, W.A.	Damage $30 000
Thunderstorm		Perth	200 houses damaged
Tornado		Adelaide	2 houses unroofed
Strong winds		Port Macquarie	Damage $400 000
Hail		Hobart	Damage $400 000 to apple crop
Frosts		Murray-Mallee, S.A. Mallee, Wimmera, Vic.	Extensive damage to cereal and vegetable crops
TC Althea	1971–72	Townsville	3 deaths, damage $50+ million
TC Daisy		Bundaberg-Maryborough	2 deaths, beach erosion, flood damage
Thunderstorm		Melbourne	Record 1 h rain: 78·5 mm, wettest month ever: 238 mm, estimated release 100 000 t water over 1 km^2 of city
Tornadoes Thunderstorms		South-east Queensland	Many houses damaged/destroyed, hail up to 100 mm diameter
TC Una	1973–74	Townsville-Rockhampton	Flooding in coastal streams
TC Wanda		Brisbane-Ipswich	Floods: 10 000 homes damaged/destroyed, 9000 people evacuated, damage $100 million
TC Zoe		South-east Queensland/ North-east N.S.W.	Gales, flooding in coastal streams
Extratropical storm		Queensland/N.S.W. coasts	2 large vessels driven ashore, 6 deaths, several yachts wrecked, extensive coastal erosion
Thunderstorm Tornado		Brisbane	Damage $2+ million
Thunderstorm	1975–76	Armidale	Damage $1 million including 500 hail-damaged vehicles
TC David		Yeppoon, Queensland	Houses unroofed, some demolished, flood damage $6 million
TC Joan		Port Hedland, W.A.	85% town damaged, 100s buildings unroofed, major flooding, damage, $20+ million
Thunderstorm		Toowoomba	Damage $12–15 million
Tornado		Perth	Damage $150 000 to state forest
TC Ted	1976–77	Mornington Is./Burketown	1050 people lost homes, damage $8 million
TC Otto		Ingham-Tully	Serious damage to sugar cane and property

lightning which involve the sudden impact of very large amounts of energy discharged in an extremely short time, and those features which become hazards only if they exceed tolerable magnitudes within or beyond certain time limits. In this category can be included wind hazards associated with extra-tropical low pressure systems, heatwaves, snow, large amounts of rain, frosts, and droughts.

The following sections give some examples of the range and variability of atmospheric hazards and extremes and of their impact, but a comprehensive coverage is beyond the scope of this book.

4.2 TROPICAL CYCLONES

Tropical cyclones (hurricanes in the north Atlantic, typhoons in the north Pacific) can be the most dangerous and deadly storms on earth (Fig. 4.1). They are usually very mobile and relatively unpredictable, and are accompanied by high velocity wind gusts, sustained high winds, torrential rains and storm surges (Fig. 4.2). Some noteworthy tropical cyclone disasters are listed in Table 4.2. Sea action and flood rank ahead of wind as causes of death and destruction by tropical cyclones. When a tropical cyclone approaches or crosses a coastline, the very low atmospheric pressure and the stress of the strong winds on the sea surface produce a rise in sea level above the normal tide level. This is the storm surge, and its effect may be amplified by the shape of the sea bed and the coastline. If the storm surge accompanies a high tide, inundation of coastal areas, major damage and loss of life can occur.

Table 4.2
SOME NOTEWORTHY TROPICAL CYCLONE DISASTERS
(after Southern, 1976)

Year	Location	Deaths
1970	East Pakistan	300 000
1737	India	300 000
1881	China	300 000
1923	Japan	250 000
1897	East Pakistan	175 000
1976	East Pakistan	100 000
1977	India	55 000
1864	India	50 000
1833	India	50 000
1822	East Pakistan	40 000
1780	Antilles	22 000
1839	India	20 000
1789	India	20 000
1965	East Pakistan	19 279
1963	East Pakistan	11 468
1963	Cuba-Haiti	7 196
1900	Texas	6 000
1960	East Pakistan	5 149
1960	Japan	5 000
1974	Darwin	49

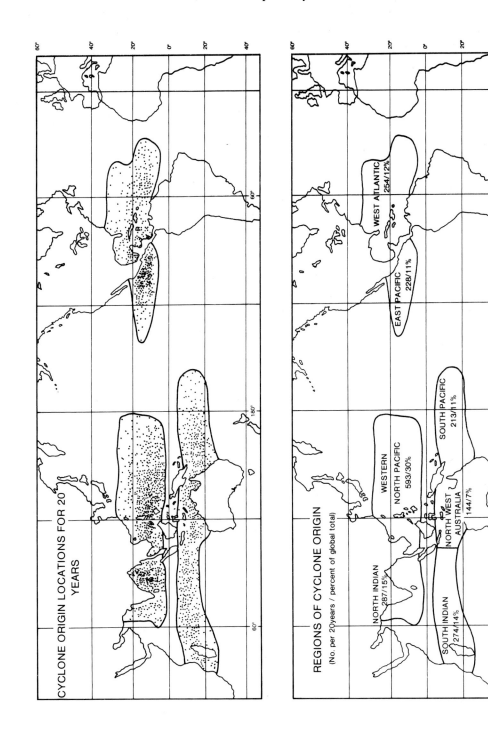

FIG. 4.1. REGIONS AND ORIGIN LOCATIONS OF TROPICAL CYCLONE GENESIS DURING THE PERIOD 1952–71 (After Gray, 1975; reproduced with the permission of Dr W. Gray).

The most recent of the events included in Table 4.2 was the tropical cyclone of November 1977 which struck the Andhra Pradesh coast of India. The cyclone generated three storm surges, the last and biggest being 6 m high, moving at 175 km.h^{-1} along a 240 km front. The last surge raced 20 km inland, killed 55 000, destroyed the homes of two million people, ruined 1·2 million hectares of crops and rendered most of the land barren (because of salt) for up to three years. The northern Bay of Bengal suffers a particularly serious storm surge problem, because of a unique combination of large astronomical tides, a funnelling coast configuration, low flat terrain and frequent severe tropical storms. The average annual frequency of tropical depressions (winds less than 63 km.h^{-1}) is twelve to thirteen, of which five strengthen and become cyclonic storms with winds greater than 63 km.h^{-1}. Most cyclonic storms remain weak and offer no serious threat, with an average of one severe cyclone about every five years (Frank and Husain, 1971).

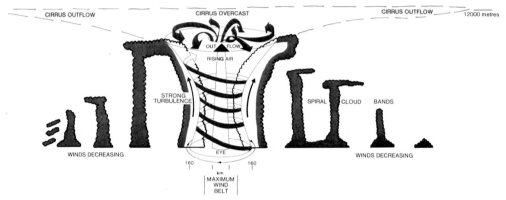

FIG. 4.2. TYPICAL FEATURES OF A TROPICAL CYCLONE

The storm which hit East Pakistan in November 1970, although only of moderate strength, may have been one of the most deadly, if not the deadliest, ever to devastate a coastal area. The occurrence of the tropical cyclone coincided with a high tide to produce a storm surge of over 6 m. About 300 000 people were killed; nearly 90 per cent of the marine fishermen suffered heavy losses; and it was estimated that about 65 per cent of the total annual fishing capacity of the coastal region of East Pakistan was destroyed (Frank and Husain, 1971).

All of the tropical cyclones listed in Table 4.2 were clearly many times more devastating than cyclone Tracy which hit Darwin on Christmas day 1974 and claimed forty-nine lives, with another sixteen posted as missing-at-sea. By global standards cyclone Tracy was a very small but intense storm, the diameter of the area of gale force winds (> 63 km.h^{-1}) being only about 100 km. Winds in the Darwin area were destructive over a path about 28 km wide, but the storm surge effect was limited because the cyclone occurred during a period of neap tides. Some north Pacific typhoons have had gale force winds up to 750 km from their centres, while the Andhra Pradesh storm of November 1977 produced gales along a 600 km stretch of the Indian coast. Maximum sustained wind speeds with this tropical cyclone exceeded 180 km.h^{-1} with gusts to 234 km.h^{-1}.

The recorded surface wind speed with cyclone Tracy indicated gusts between 217 and 240 km.h^{-1}, corresponding to estimated mean wind speeds of 140 to 150 km.h^{-1} (Bureau of Meteorology, 1977). Larger tropical cyclones, which have developed with a longer travel over tropical seas, can attain hurricane force winds (>120 km.h^{-1}) up to 100 km from the storm centre. The greatest wind velocities are usually associated with the maximum wind ring around the eye of the centre of the storm. The central eye (diameter usually 20–30 km, Tracy 12 km, Andhra Pradesh 60 km) is accompanied by a lull of from a few minutes to an hour or two, in which calm or light winds prevail. After passage of the eye strong winds return from a different direction.

In the case of cyclone Tracy the eye of the storm passed directly over Darwin (Fig. 4.3). Eighty per cent of the buildings in the city of 45 000 people were damaged, and the city was left without light, power, water, sanitation and communications. The intensity of the cyclone destroyed, or extensively damaged, all radio mast structures and associated buildings and for several days communications between Darwin and the rest of Australia were unreliable. The major factor contributing to the loss of communications was water damage from rain entering buildings, housing exchanges and transmitters.

There is little doubt that the relatively very small loss of life in Darwin can be attributed to the fact that the cyclone was accompanied by a restricted storm surge effect; major damage was produced by high winds and torrential rain. Nevertheless, the problems encountered in Darwin, serve to highlight the magnitude of the tropical cyclone danger in much more densely populated areas of developing rather than developed countries. In addition most tropical cyclone statistics mask the fact that the greatest potential destruction is at sea, where meteorological conditions are most favourable to maximum energy input and where surface friction is least. This means that features such as coral reefs, sand cays and the vegetation of small exposed islands are especially vulnerable. Shipping and oil rigs (which cannot run for shelter) are particularly at risk. Also, vertical circulation within a tropical cyclone is strongly turbulent to heights of several kilometres, posing considerable dangers to aircraft.

Consideration of the possible disastrous effects of tropical cyclones raises the question of whether people and their governments must passively accept that every so often a tropical cyclone will occur and hope that the effects will be small, or whether they can take well planned measures which might prevent, or at least reduce, losses of lives and damage (ESCAP, LRCS and WMO, 1977). The answer seems to be that carefully planned measures functioning through an efficient organization are well worth the effort.

In some countries, notably the USA and Japan, and to a lesser degree Australia, highly developed systems already exist for disaster prevention and preparedness. Extensive long-term measures such as flood control, land use control, zoning and building codes have been introduced or are under consideration. Australia is still digesting the lessons of events such as cyclones Ada (1970), Althea (1971), Tracy and Wanda (1974), but there is still a common feeling that action is slow, with public and bureaucratic apathy yet to be fully overcome.

Data from the USA (Fig. 4.4) show clearly how effective disaster preparedness and prevention programmes, incorporating adequate warning systems, can substantially reduce the death toll from tropical cyclones. The point to emphasize is that while improvements in the hurricane warning service and community preparedness programmes

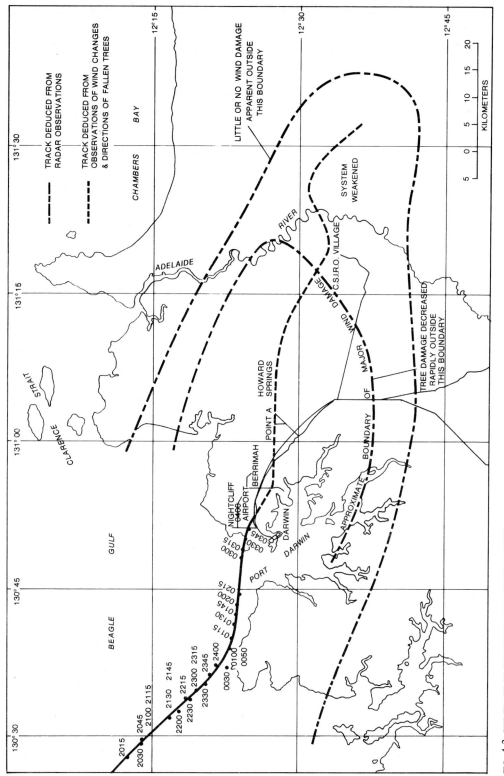

FIG. 4.3. THE TRACK OF CYCLONE TRACY ACROSS DARWIN
(After Bureau of Meteorology, 1977; reproduced with the permission of Australian Bureau of Meteorology, Department of Science and the Environment).

TRENDS of LOSSES from HURRICANES in THE UNITED STATES

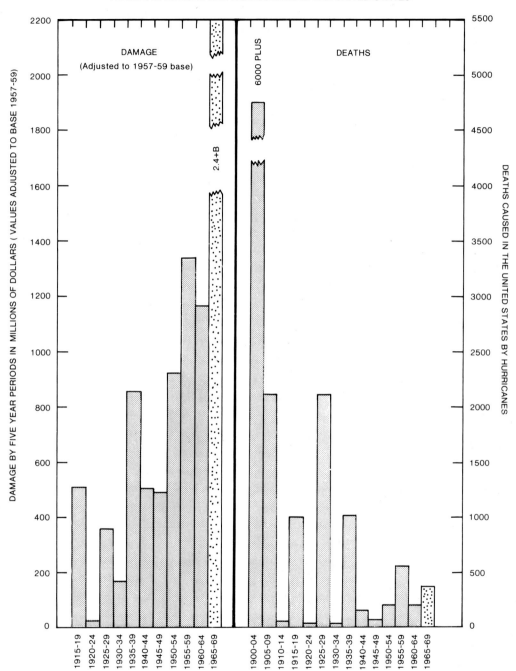

FIG. 4.4. TRENDS IN LOSSES FROM HURRICANES IN THE USA
(After ESCAP, LRCS and WMO, 1977; reproduced with the permission of the World Meteorological Organization).

have decreased the death toll, the average annual damage, adjusted for the effects of inflation, was nine times as great in the decade 1965–74 as in the period 1915–26. This is largely a reflection of the increasing numbers and volume of man-made structures erected in vulnerable areas.

An important element in community preparedness to cope with tropical cyclones is that of design standards for buildings. Society tends to place considerable responsibility on the engineering profession for ensuring that the impact of natural hazards when they do occur is not too serious. Most building damage is caused by the effects of winds on building types that are not fully engineered, such as domestic houses and small low-rise industrial structures. Cyclone Tracy produced 3-second wind gusts acting on buildings estimated at 55 m.s^{-1} (200 km.h^{-1}). Engineered structures generally performed well, but domestic construction fared badly. Damaged buildings included those which were considered to conform to current standards of good building practice for tropical cyclone regions. The damage was mainly due to inherent deficiencies in the design of traditional domestic structures. The fact is, however, that the various traditional forms of Australian house are not easily modified to resist wind forces encountered in tropical cyclones, so it may be necessary to develop a new type of house that is not prohibitively expensive to construct (Leicester and Reardon, 1976).

The warning system is a primary feature of organization for disaster preparedness. Key elements in a warning system are the efficiency of the dissemination process and the reaction of the community, as well as the accuracy of the warning itself. Education is vital to ensure that people understand the warnings and know how to respond, and to ensure that the public is aware of the necessity to heed warnings. In their report on the East Pakistan cyclone of November 1970, Frank and Husain (1971) emphasized the problems of providing cyclone warnings, pointing in particular to the considerable apathy in the population. There had been many false warnings in the past because of lack of facilities to distinguish between killer and non-killer cyclones, the overwarning inducing apathy. It was estimated that over 90 per cent of people in the disaster area knew about the storm, yet less than 1 per cent sought refuge in more substantial buildings. Most residents had never experienced a storm surge like that of November 1970 and thus felt no urgency to leave their homes (Table 4.3).

A similar situation seems to have occurred in Darwin with cyclone Tracy. Three weeks before Tracy, cyclone Selma had seriously threatened Darwin, but turned back to sea after producing only fringe effects in the city. This contributed to public complacency, although

Table 4.3
ESTIMATED DAMAGE IN THE 12 NOVEMBER 1970 CYCLONE IN EAST PAKISTAN
(after Frank and Husain, 1971)

Population affected	4·7 million
Crop losses	US$ 63 million
Loss of cattle	280 000
Loss of poultry	500 000
Houses damaged	400 000
Schools damaged	3 500
Fishing boats destroyed (marine)	9 000
Fishing boats destroyed (inland waters)	90 000

records indicate several near misses for Darwin and at least six tropical cyclones which have seriously affected the Darwin community. The last severe event was, however, in 1937 and was remembered by relatively few residents in 1974. In addition Christmas Eve was hardly an ideal time to muster enthusiasm for positive action by organizations or the public. Falls (1978) has suggested that the potential benefits of technically good predictions were not realized. The average 12 hour forecast position error in cyclone warnings for Tracy was only 42 km (Bureau of Meteorology, 1977), much smaller than average performance in Australia or the USA.

The general situation has been summarized very well by Southern (1976). He viewed an efficient tropical cyclone warning system as an excellent example of science in the service of humanity; but the benefits of the undoubted improvements in detection technology and in behaviour prediction are lost without corresponding progress in man's ability to utilize the information in planning, organizing and acting for his protection and convenience.

4.3 SEVERE LOCAL STORMS

Tornadoes and thunderstorms are the main forms of this category of atmospheric hazards. Their formation generally relates to conditions favouring the release of instability.

Tornadoes are among the smallest but most destructive atmospheric circulation features. A tornado is commonly defined as a violently rotating column of air, usually visible as a pendant funnel cloud growing downwards from the base of a thunderstorm. Many tornadoes are not directly associated with the giant cumulonimbus clouds of the parent, severe thunderstorm but with the flanking lines of cumulus clouds that merge into the rotating cumulonimbus.

The funnels of most tornadoes have widths of only 100 to 1000 m at ground level, although typical New Zealand tornadoes have been reported to have damage paths only 10 to 30 m wide (Tomlinson and Nicol, 1976). Pressures and windspeeds in a tornado vortex are very difficult to measure. A pressure decrease of as much as 200 mbar is theoretically possible, but the greatest recorded decrease is only about 40 mbar. Windspeeds may reach nearly 500 km.h^{-1}, but most tornadoes do not have winds exceeding 250 km.h^{-1}. Tornadoes exhibit a considerable range of intensities, sizes and durations. They vary from narrow, ropelike funnels having lifetimes of a few minutes to giant, very intense, long-track tornadoes (Golden, 1976). Fujita (1973) reported twenty-three tornadoes in the USA in 1972 with path lengths no longer than 46 m, yet some have been known to travel up to 400 km. The average length of a tornado track is about 15 to 20 km, with travel generally being in the same direction as the clouds to which the vortex is attached.

Damage characteristics associated with tornadoes include windows and walls bursting outwards, tops screwed or snapped off trees, severe structural damage, and relatively localized patterns of destruction. The dynamic force of the winds and the reduced pressure at the centre of a tornado can lead to the explosive collapse of structures in the vortex path. In addition, tornado-generated missiles present engineers with considerable design problems. In tornado-prone areas of the USA where protection of people in buildings is important, the tornado missile is the controlling design factor. Buildings such as hospitals, fire stations, and emergency operating centres, where critical functions must be maintained, are particularly susceptible to damage from missiles. Missile protection for nuclear power plant facilities is costing millions of dollars in additional construction costs (McDonald, 1976). At present there is still considerable uncertainty concerning what missiles will fly in a tornado of given intensity and the maximum velocity they may attain (Table 4.4).

Table 4.4
USES OF TORNADO TECHNOLOGY
(after Minor, 1976)

Current and Potential Areas of Use		Product of Use
Nuclear Facilities	Nuclear power plants	Safe, economical facilities design
	Nuclear materials processing	Safe, economical facilities design
Life Safety	Weather advisory systems	Effective watch/warning systems
	Shelters in homes, schools, and public buildings	Safe, accessible, economical shelters
Hazard Analysis	Risk analysis	Perspective on relative importances of hazards
	Facilities evaluation	Designs for hazards reduction
Building Design	Economics of design	Economical wind resistant construction
	Protection of critical functions	Preservation of essential services
Housing	Wind resistant design	Minimum life-cycle cost
	Shelters	Low-cost, in-home shelters
Social-Economic Systems	Human behaviour	Minimize anxiety and disruption of life
	Cost effectiveness	Guidelines for coping with threat
Planning	Community planning	Integrated preparedness, advisory, and action system
	Organizational Planning	Preparedness for natural hazards

Tornado missiles range in size from gravel to motor vehicles as large as semi-trailer trucks. The most common missiles are pieces of timber from destroyed residences and other timber constructions. Pieces of timber range from splinters to laminated beams or arches used as roof supports. Sometimes a whole section of roof may be ripped off and carried a long distance by tornadic winds. Small outdoor equipment, and larger objects such as utility poles have been observed to become tornado missiles. A tornado at Lubbock, Texas, in 1970 was responsible for moving a long cylindrical fertilizer tank ($3 \cdot 35$ m $\times$ $12 \cdot 5$ m) weighing over 11 t about $1 \cdot 21$ km from its original location. It was calculated that the windspeed necessary to roll the tank out of its saddle support was $81 \cdot 8$ m.s^{-1}. Three forty-passenger school buses apparently became airborne in a tornado at McComb, Mississippi, in 1974 (McDonald, 1976).

In the USA tornadoes cause about US$ 100 million damage and 150 fatalities annually (Abbey, 1976). The considerable year-to-year variation is exemplified by the 148 tornadoes which occurred on 3–4 April, 1974, causing over US$ 400 million damage and killing more than 300 people. Court (1976) acknowledged that tornadoes are important for the people they kill and the damage they do, but pointed out that damage is usually reported in monetary terms instead of the more useful measure, unaffected by inflation, of devastated area.

Many tornado funnel clouds are not seen, perhaps because they occur at night or in poor visibility, perhaps because they occur in remote or relatively unpopulated areas. Records of tornadoes in Australia, for example, are of very variable reliability and precision. It is considered possible that the frequency of occurrences in some parts of Australia approaches that of the USA. Reports of tornado occurrences tend to give the impression that they are, indeed, most common in the USA. This may or may not be an accurate impression, since the frequency of reports of such short-lived, small-scale phenomena is probably related to population density. Since 1950 every year in the USA there has been an average of 662 tornadoes, resulting in 114 deaths. In 1977 there were 850 tornadoes, killing 43 people. In the first seven months of 1978 the number of reported tornadoes in the USA was about 700, much higher than average. Yet the number of deaths was only 39, less than half those usually reported. Reported tornadoes in England during the first seven months of 1978 totalled 22, which in terms of tornadoes per unit area is nearly twice as many as for the USA. The annual frequency of damaging tornadoes in Britain has been put at between 30 and 50.

It is still difficult to explain tornado formation, why they form only in storm clouds and never in still air, or why few of the storm clouds that should spawn them actually do. New radar techniques developed in the USA have enabled scientists to identify tornado signatures in thunder clouds. Observations using these techniques have indicated that the vortex begins to form in the middle levels of the cloud at 4–5 km altitude. The tail then descends to the ground about half an hour later. Recent Australian research using mathematical modelling of tornadoes is producing this sort of picture and seems to be getting close to a solution to some of the problems of tornado formation (Leslie and Smith, 1978).

Pearson (1976a) notes that the forecaster is looking for a synoptic situation combining seven elements: mass convergence near the surface, mass divergence aloft, a buoyant air mass, wind shear in the vertical, moist air mass in the lower layers, a trigger mechanism, and surface cyclogenesis. Although none of these parameters is uncommon, their juxtaposition is very uncommon. In most regions experiencing tornadoes the data network is inadequate to be ideal for the measurement of such small-scale weather events. About the best that can be done in terms of forecasting is to specify broad areas over which atmospheric conditions seem favourable for tornado occurrence. Thus, meteorologists are looking for a link between the large-scale weather patterns and the thunderstorm scale of activity.

The unpredictability and ferocity of tornadoes make protection and preparedness very difficult. Nevertheless, in the period that tornado forecast and warning services have been in operation in the USA, the annual average number of fatalities seems to have decreased markedly. Cressman (1969) reported that from 1916 to 1952 the average number of deaths due to tornadoes was over 200, whereas from 1953 to 1968 this was cut to about 120. At the same time the number of tornadoes reported each year quadrupled to over 600. This increase reflects the increased exposure of people because of a growing population, as well as improved reporting procedures. Cressman attributed the drop in fatalities to the efficiency of the tornado warning system, although even the best systems cannot prevent considerable loss of life if a tornado first touches down in a populated area.

Two methods of detection provide the mainstays of the operational detection and tracking system in the USA: weather surveillance radar and prompt reports of visual sightings by competent observers. The ability to detect the existence of a likely tornado-bearing storm with radar varies markedly, with the size and intensity of the associated tornado and with the distance of the storm from the radar site (Lemon et al., 1977). Usually prompt reports of visual sightings provide the first indications of the onset of tornado activity, perhaps confirming suspicious radar echoes (Darkow, 1976; Muench, 1976).

Tornado warnings are issued only when a tornado has actually been sighted or when its existence or development have been strongly indicated by radar. Differences in community interest, resources and communication capabilities markedly affect local distribution and dissemination of warnings. Responsibility for establishing and maintaining effective warning systems falls on to many groups: local National Weather Service officers, local civil authorities, radio and television stations, Civil Defense Preparedness groups, law enforcement agencies and others. Many of the problems associated with warning dissemination involve economic, sociological and political considerations.

Thunderstorms are a much more common and widespread, but generally less destructive phenomenon than tornadoes. As separate cells or as organized line squalls, thunderstorms develop cold downdraughts, bringing cold winds with high velocities, perhaps causing severe localized damage which is often compounded by intense rain or large hail and lightning. The squall of a thunderstorm can reach gusts of over 185 km.h^{-1} over 3 to 10 seconds. Usually only small areas are affected, with an individual storm travelling something like 60 km. A link with thermal conditions at the surface is shown by a tendency for thunderstorms to develop in the afternoon or early evening.

In summary, severe local storm hazards, of whatever type, are widespread, relatively unpredictable, seemingly impossible to prevent and often costly in terms of lives lost and damage caused to property (Atlas, 1976; Newton, et al., 1978).

4.4 CUMULATIVE HAZARDS

Many atmospheric disasters are due to an accumulation of events, which singly would not be hazardous. One dry day or even one dry year does not necessarily constitute a drought, but a succession of abnormally dry years can have disastrous effects on the environment and on man. Similarly, hot days are common in many parts of the world, but a succession of many very hot and very dry days can prove lethal (e.g. Tout, 1978). Any succession of otherwise quite common atmospheric conditions may produce a winter of markedly higher than average snowfall, or a summer of drought proportions, perhaps presenting some hazard to the well-being of man.

Most people are reasonably well aware when a drought situation exists, but it is very difficult to find an overall acceptable definition of drought. It clearly involves a shortage of water, but can really be defined only in terms of a particular need. Different types of resource use have different water deficiency thresholds. Drought is as much a problem of demand for water as a reduction in the natural supply. The most common view of drought is of rainfall deficiency, but the links between rainfall and the water which becomes available to meet a demand are complex. Therefore, definition of a drought relates not only to water needs but also to the complex set of factors involved to supply that need through the hydrological cycle.

An agriculturalist or pastoralist in a relatively dry region may be accustomed to perhaps 200 mm of rain in a year, but this would be a disastrous drought to a farmer usually receiving falls of 500 to 800 mm per year. As Hounam et al. (1975) point out: 'The agriculturalist or pastoralist, especially in the drier regions, has assessed the nature of local rainfall and, through years of long and sometimes bitter experience, has learnt to adapt his operations to the rainfall characteristics of the area'. In other words drought is related to the failure of the usual rains at a particular time, since most activities using water will be geared to that which is normally available.

Heathcote (1976) has identified further problems related to the definition and perception of drought: that the measurement of the significance of a moisture deficiency is in terms of user expectations, which vary with the type of resource use and between similar resource users with different experiences of past conditions; and that the reporting of drought occurrence may be over or under-estimated since the material well-being of the reporter may be affected. In Australia, for example, the traditional policies of drought relief mean that it may be in the interest of the resource manager to claim drought impacts whenever possible, since his losses may be alleviated from public funds.

Despite the obvious limitations of rainfall as an indicator of water available for various purposes, it is commonly used on its own or in combination with other parameters to identify drought. Rainfall alone is used to denote a drought situation in terms of when it is less than a specified amount, perhaps a proportion of the normal, for a period of time. Many definitions derived in Europe or the northeast USA, where agriculture and other activities are geared to rainfall received at fairly frequent intervals, refer to relatively short time periods, perhaps less than a month (Hounam et al., 1975). These are of limited use in seasonal rainfall areas such as occur over much of the intertropical zone, where agricultural farming is geared to a distinct seasonal water regime. Rainfall values, however, have limitations as drought indicators, so many definitions and indices incorporate other parameters such as evaporation, humidity, air temperature, solar radiation, wind, soil moisture, streamflow and plant condition (Table 4.5).

Accepting the problems of definition, there are several aspects of drought impact which should be considered. These include the ecological, economic, demographic and political significance of drought. Long term impacts of drought on plant and animal life have been generally ignored. Many questions seem to be unanswered concerning the role of drought in an ecosystem. It is not clear, for example, whether drought destroys the character of an ecosystem or whether it is a necessary part of that ecosystem, perhaps providing a natural control on plant and animal populations. In addition, drought may have different ecological significance in different ecosystems, possibly being most significant in zones transitional between humid and arid ecosystems (Gentilli, 1971b).

The impact of drought upon human activities is usually described in terms of reduced water supplies and economic losses, although evaluation of such impacts is complicated. Many factors need to be taken into consideration when attempting to calculate economic losses because of drought. These include the fact that shortfalls in expected crop or livestock yields tend to inflate the value of actual yields upon which the value of the lost

Table 4.5
DROUGHT DEFINITIONS BASED ON PRECIPITATION
(after Hounam et al., 1975)

Author	Description of Drought Definition
Bates, C. G. (1935)	Annual precipitation 75 per cent or less of normal when monthly precipitation is 60 per cent or less of normal
British Rainfall Organization (1936)	Absolute drought: at least 15 consecutive days with less than 0·01 in per day. Partial drought: at least 29 days when mean rainfall does not exceed 0·01 in. Dry spell: 15 consecutive days with less than 0·04 in. per day
Hoyt, J. C. (1936)	Any amount of rain less than 85 per cent of normal
Baldin-Wiseman, W. R. (1941)	Engineers' drought in Australia: 3 or more consecutive months with deficit of 50 per cent from normal
Tennessee Valley Authority (1941)	No interval of 21 days receiving precipitation greater than one third normal
Blumenstock, G. (1942)	Less than 0·10 in during 48 hours
Conrad, V. A. (1944)	Period of 20 or more consecutive days without 0·25 in precipitation in 24 hours (during March to September)
Ramdas, D. A. (1950)	When rainfall for a week is half normal or less
Fitzpatrick, E. A. (1953)	Period terminated by at least 0·25 in during any 48 hours
Foley, J. C. (1957)	Cumulated departures from monthly average rainfall. Divided values by average annual rainfall to give 'index of severity' and improve comparability between stations
Henry, A. J. (1960)	21 days or more when rainfall 30 per cent or less of average Extreme drought when rainfall less than 10 per cent of average for 21 days or more
Gibbs, W. J. and Maher, J. V. (1967)	Deciles of rainfall demonstrating temporal distribution Mapped areas of first decile range rainfalls coincide approximately with areas of severe drought.

production is calculated, although there is no guarantee that the lost production could have been sold anyway (Heathcote, 1976).

The 1975–76 drought in the UK provides a good recent example of the impacts of water shortages in an area not normally noted for such problems on any major scale. The twelve month period starting 1 May 1975 was the driest such period in the UK in the period of record, with less than 60 per cent of normal precipitation in parts of the Midlands and south-west England (Perry, 1976). 1975 as a whole was the fifth driest year this century and there had been other very dry recent years in many areas, with 1973 even drier on average over the UK than 1975. The four months from May to August 1976 were also very dry, with estimates suggesting that the summer was one of Britain's two hottest of the last three centuries. The synoptic conditions responsible for this prolonged dry spell have been discussed by several authors in considerable depth and will not be examined here (e.g. Mortimore, 1976; Perry, 1976; Green, 1977; Miles, 1977; Ratcliffe, 1977; Kelly and Wright, 1978). The impact of the prolonged drought was progressive. By the end of the 1975 summer some water restrictions were already in operation and reservoir levels were beginning to fall rapidly due to frequent high temperatures. The mostly wet September 1975 did little to ease the situation because most of the rain replenished soil moisture in the parched ground rather than producing runoff to fill reservoirs. The water situation continued to deteriorate through the 1975–76 winter and into summer 1976. Bans and restrictions were in operation over most of England and Wales, and in Gwent and parts of south and mid-Glamorgan water supplies to domestic consumers were withdrawn for up to 17 hours per day (Mortimore, 1976).

Mortimore reported the effects on industry to be not too serious, although voluntary reductions in use of water were nationwide. In south-east Wales firms were told to cut consumption by 50 per cent and in the Midlands by 40 per cent. The agricultural community was generally much harder hit. A study by Roy et al. (1978) has pointed to some of the agricultural effects of the 1975–76 drought. Even after the dry winter of 1975–76, soils under short-rooted vegetation were at field capacity in Febuary 1976 in most areas. In the following main part of the growing season (April to August) soil moisture deficits increased with the lack of rainfall (Fig. 4.5). While the mild winter probably favoured the

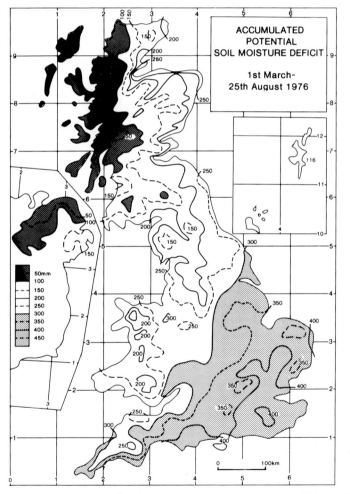

FIG. 4.5. POTENTIAL SOIL MOISTURE DEFICITS (mm) ACCUMULATED BETWEEN 1 MARCH 1976 AND 25 AUGUST 1976 (After Roy et al., 1978; reproduced with the permission of the Royal Meteorological Society).

carry over and early development of cereal diseases in winter wheat, the dry weather of early summer helped to produce a remarkably disease-free year. This to a certain extent helped to alleviate the reduction of output because of the drought. In the opinion of Roy et al. the overall decreases in yields of cereal crops between 1974 and 1976 were not very large considering the severity of the drought and the extremely high mid-summer temperatures. Potato crops in England and Wales showed very low yields (Table 4.6) and, paradoxically, the wet autumn made their harvesting very difficult. Grass growth was severely restricted for much of the summer, but recovered very rapidly when the rain finally came.

Table 4.6
ANNUAL ESTIMATES OF AVERAGE CROP YIELDS
$(t.ha^{-1})$ (after Roy et al., 1978)

	1974	1975	1976 (provisional)	Percentage decrease 1974 to 1976
England and Wales				
Wheat	4·94	4·30	3·85	22
Barley	3·95	3·40	3·46	12
Oats	3·88	3·45	3·42	12
Potatoes—early	18·8	14·1	16·3	13
—main crop	33·9	22·1	20·4	40
Scotland				
Wheat	5·75	5·58	4·98	13
Barley	4·99	4·79	4·10	18
Oats	3·66	3·45	3·58	2
Potatoes—early	24·1	19·3	19·4	20
—main crop	30·9	26·6	26·2	15

Although it might be suggested that agricultural losses were not as severe as might have been expected, the cost of lost crops exceeded 500 million pounds. Many local authorities incurred great expense, perhaps up to one million pounds each, through drought alleviation measures such as stand-pipes, extra pumping equipment, new pipelines, the search for additional water supplies, and massive publicity campaigns. Part of the national government's emergency programme even included the appointment of a Drought Minister, who gave priority in the queue for water to agriculture and industry, with domestic users last in line.

In France an unexpected consequence of the 1975–76 drought was the introduction of a special tax to raise revenue for the support of agriculturalists who had suffered most. The amount levied depended upon the taxable income of the individual.

Drought is a much more common feature of the Australian scene, although it still tends to be regarded as an unfortunate and irregular abnormality of the environment. This view may be a little unrealistic, since much of interior Australia is marginal anyway in relation to the availability of water, and it might be more appropriate to consider drought as part of the normal chain of events. Gibbs and Maher (1967) presented a comprehensive summary of droughts in Australia, illustrating that they are both relatively frequent and widespread.

Nevertheless, it is difficult to document precisely the influence of drought on patterns of land use. Federal government policies on land settlement and price supports, world market prices and an ageing rural population structure are major influences on land use in Australia, but there is evidence that drought has often played a major role in promoting changes in land management and resource use (Heathcote, 1976).

The massive drought of the 1890s is a good example of the lack of understanding of a new environment on the part of recent settlers. Although each new district had already experienced very dry years and some droughts, most land-holders considered droughts to be abnormal. There was widespread optimism about the climate of much of Australia and particularly of the south-eastern quarter which produced most of the rural wealth. Thousands of farmers before 1890 believed that the climate was on their side, and in South Australia, the leading wheat colony, there was even a belief that rain followed the plough. This view was officially blessed by the Minister for Agriculture in South Australia in the mid 1870s (Blainey, 1977). Notions that ploughing and tree planting could bring rain were widespread, enticing farmers into marginal areas. The overall picture during the 1870s and 1880s was of rapid growth of rural enterprises and their associated infrastructures.

In the 1890s, however, the worst financial disaster in the country's history was followed by the worst drought so far recorded. The first year of the drought was probably 1895 and it ended in 1903. The impacts of the prolonged drought were many and varied. The number of sheep estimated to live in Australia fell from 106 million in 1891 to only 54 million in 1902, and cattle numbers had almost halved to about 7 million. Dust storms were common. The area of land ploughed for wheat in south-eastern Australia was almost six times as large in 1900 as in 1866. This vast extent of tilled soil no longer compacted by moisture, crumbled into powder, and was blown away by the winds. Many fences were submerged by the drifting soil and stretches of railway line were buried. On 21 November 1902, so much soil was blown from the interior that Melbourne was drenched with dust, and the afternoon sun was almost hidden by the dust in the air. The Bureau of Meteorology reported that in inland towns that afternoon it was almost as dark as night (Blainey, 1977). This was, unfortunately, not an isolated drought. General droughts occurred again in 1911–16 and 1919–20, so that the early part of this century was not at all favourable to agriculture over most of south-eastern Australia. The overall impact of the prolonged dry spell was to bring about a retreat of agricultural frontiers back towards the coast.

Periodic droughts of varying intensity have been experienced throughout Australia since the first quarter of the century. Two periods of widespread drought in the 1960s, 1965–66 and 1967–68, resulted in a marked fall-off in rural production. These two periods were among the driest Australia has experienced, but generally favourable influences on rural production in the early 1960s made it easier to cope with drought than it had been in earlier periods. An examination of the main effects on the economy indicates that while farm gross national product suffered a serious setback of about 20 per cent in each of the two years most affected, it recovered to establish new records in the years immediately following (McIntyre, 1973). Most of the effects of drought were short-term, in the years when drought was at its worst, but some longer-term effects may have resulted from loss of capital which reduced output in subsequent years. Clearly, also, many individual producers would have suffered negative incomes.

It seems that drought will decline in significance as a factor in the economies of developed countries. In countries such as Australia the rural sector is contributing a declining proportion of national output, and general ability to cope with drought is increasing.

4.5 CLIMATIC CHANGE AND WORLD FOOD PRODUCTION

A major current concern about possible climatic change centres upon global food supply. The size and rate of growth of the world's population have put increasing pressures on the cultivation of climatically marginal land. There is a great reliance on assured levels of crop yields. Weather-induced reductions in yield, especially in the major food-producing areas, have become of crucial political, social and economic importance. In 1972, for example, drought caused the USSR to purchase over 18 million tonnes of grain from the USA and more than 12 million tonnes in 1975, with repercussions on domestic food prices and balance of payments (Bryson and Murray, 1977). In 1974 world grain stocks were at their lowest for 26 years, and a combination of floods in Canada and drought in the USA forced the price of wheat and corn up by one third and soya beans up nearly two thirds (Gribbin, 1976b). During these early years of the 1970s, there have also been droughts in the Sahel and Central America, severe frost damage to coffee crops in Brazil, and many other regional climatic anomalies.

The short-term variability of climate, associated with such events, is superimposed on longer term trends over a variety of time periods. It is becoming increasingly important to distinguish isolated events producing the occasional bad year from any developing trends towards establishment of a new set of normal conditions. According to Laur (1976), two facts have become clear as a result of the developing food crisis: climatic variation must be taken into account when future food supplies are estimated, and agricultural production will be used as a political tool on international bargaining tables. This points to the need for nations to maintain strategic reserves of food grains to offset years of poor agricultural production.

Over the past quarter of a century up to the early 1970s agricultural output in the USA has shown a steady increase based on technological progress, improved methods and machinery and generally favourable weather conditions. Major food producing areas in South and South-east Asia benefited from the 'Green Revolution' with its miracle wheat and rice, which temporarily offset the decline in world food grain reserves (Fig. 4.6). Thus optimism for the future was the keynote, but it ignored the possibility of a variable climate. Analysis of long-term yield trends for wheat, maize and soya beans in the USA over an 80-year period (NOAA, 1973) suggested that technological change accounted for 70 to 80 per cent of yield variance over time, and weather variability for only 10 to 20 per cent. An analysis of Australian crop yields showed a contrasting 60 to 80 per cent of the variance due to weather variability (Russell, 1973). The difference may be attributed to the fact that a much higher proportion of Australia's crops are grown in climatically marginal areas.

One approach to the problem of predicting the possible impact of changing climate is through the use of historical data and crop–climate relationships to build models of crop productivity under varying climatic conditions. Such a simulation model of sheep production on subterranean clover pasture was used to examine the effects of changes in rainfall and its distribution and of changes in mean ambient temperature in localities in Western Australia (Arnold and Campbell, 1972). Economic analyses were also made, as detailed in Table 4.7 (Arnold and Bennett, 1975). At lower stocking rates the ecosystem is much less sensitive to variations in climate. The studies illustrate that pasture ecosystems are adapted

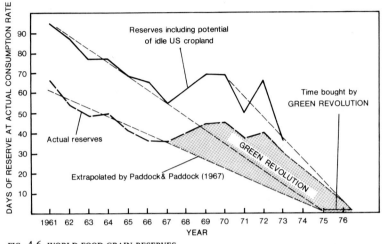

FIG. 4.6. WORLD FOOD GRAIN RESERVES
(After Bryson, 1974b; reproduced with the permission of Professor R. Bryson).

to present climate, particularly patterns of rainfall. Any changes in these patterns could have serious consequences for agricultural production in such environments.

The Climate/Food Research Group of the University of Wisconsin, Madison, devised a spring wheat model to examine the relationships between climate and wheat yield. The model considers monthly temperatures and rainfall, pre-season moisture, technological

Table 4.7
THE EFFECTS OF CLIMATE CHANGES ON PROFIT[a] AND RISK FROM SHEEP PRODUCTION AT BAKERS HILL, ESPERANCE, AND MERREDIN, WESTERN AUSTRALIA
(after Arnold and Galbraith, 1978)

| | | Average temperature | | | Rainfall | | |
		Normal	+1 °C	−1 °C	+10%	−10%	Normal
Bakers Hill	Stocking rate (sheep/ha)	14	14	14	14	14	12
	Income (A$ '000)	68·0	79·0	57·0	78·0	55·0	66·0
	Risk	16·4	10·0	35·0	11·0	32·0	10·0
Esperance	Stocking rate	10	10	10	10	10	9
	Income	49·9	52·8	45·3	54·9	47·8	48·0
	Risk	17·3	16·4	18·6	14·1	17·0	11·0
Merredin	Stocking rate	7	7	7	7	7	5
	Income	31·5	30·5	17·6	27·6	13·8	20·0
	Risk	10·5	9·0	22·4	16·2	19·4	8·0

[a] income from sales less variable costs

trends and soil type and has reproduced past yields with about 97 per cent accuracy for the USA spring wheat region. Table 4.8 shows that a positive temperature anomaly of 1 °C for each month of the growing season would only change the yield by about 7·5 per cent, but a 20 per cent rainfall reduction each month would cost another US$ 139 million in gross value. The impact on the net return of the farmer would be far greater than 7–10 per cent (Bryson, 1978).

Table 4.8
GROSS SPRING WHEAT INCOME LOSSES IN THE UNITED STATES DUE TO DEVIATIONS OF
TEMPERATURE AND PRECIPITATION FROM NORMAL (US $ millions)
(after Bryson, 1978)

| | Temperature | | Precipitation | | | |
	+1 °C	−1 °C	+1 mm	−1 mm	+20%	−20%
Preseason	—	—	+7	−8	+21	−30
April	+40	−40	+3	−3	+22	−25
May	−22	+13	+2	−2	+4	−37
June	−70	+70	+2	−2	+37	−44
July	−78	+92	—	—	−2	−2
Whole season	−131	+136	+14	−15	+82	−139

The important effect of climatic change and variability on crop production in marginal areas is well documented for Iceland. The climatic record of Iceland shows generally higher temperatures in the few centuries after Viking settlement about 1100 years ago. The coldest period was between the early seventeenth and twentieth centuries, and was characterized by frequent famines because the sensitivity of hay yield to summer temperature is about 15 per cent per degree Celsius (Bryson, 1974b). The recent decline of temperatures has apparently reduced yield nearly a quarter. Fridriksson (1973) showed that a drop of 1 °C in mean annual temperature led to a reduction in hay (dry matter) yield by one tonne per hectare. Hay is the basis of practically all land-based agriculture in Iceland, so the impact on the economy has been severe. A return to the temperatures of the last century would reduce yield by about 40 per cent (Bryson, 1978).

The world's need for food increases every year. In 1976 there were 4 billion people requiring food, and it is estimated that by the year 2000 world population will be at least 6 billion (N.A.S., 1976). The increasing demands of a growing population mean that there may be more frequent imbalances between food supply and need, leading to depletion of grain reserves, malnutrition, starvation and political unrest. Although agricultural production depends upon a variety of factors, climatic inputs provide the essential resource base (Fig. 4.7). It is clear that, even with today's technology, we are not immune to the effects of climatic variation, and we know that climate does vary.

While climate may directly influence human health, its greatest impact on man is indirect through its effect on crop and livestock production. In 1972, when the climate was particularly unfavourable for food production worldwide, millions of people starved. There is still considerable uncertainty about the nature of present climatic trends, although many of the signs point towards less favourable conditions for agriculture in some of the major producing regions. A return to conditions which are known to have occurred in the past, and which therefore can happen, could have dramatic consequences. Information is

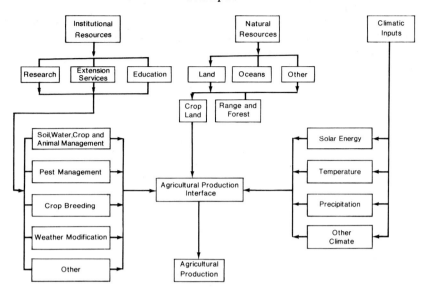

FIG. 4.7. FLOW OF RESOURCES AND INFORMATION TO SUPPORT THE FOOD PRODUCTION SYSTEM UNDER THE RESTRAINTS OF CLIMATIC FLUCTUATIONS
(After NAS, 1976; reproduced from *Climate and Food* (1976), p. 2; with the permission of the National Academy of Sciences, Washington, D.C.).

urgently needed on when and where benign or hazardous climate and weather will occur. Bryson (1978) has suggested that well-established cultures in good locations could probably cope with changes of the magnitude that may be occurring, but the picture is less hopeful in marginal situations.

4.6 IMPACT OF EXTREME EVENTS ON TRANSPORT

Since it would be impossible to consider all, or even many, aspects of the impact of extreme events on economic activity, the case of transport has been selected as a useful illustration of the nature and dimensions of the problems involved.

Many aspects of air transportation are affected directly by weather phenomena, including accidents, flight delays, airport operations, routing and fuel economy. Weather exerts a strong influence on air carriers, general aviation operations and the air traffic control system (Bromley, 1977). Data for the USA in 1975 show that over 85 per cent of total delays to aircraft for periods of at least 30 minutes were weather related (Fig. 4.8). The effect of weather on safety is demonstrated in figure 4.9, which shows that weather is seen as the most prevalent causal factor in total general aviation accidents, being responsible for 36 per cent of the total. Bromley (1977) points to the important effect of weather-related phenomena on the efficiency of the air transportation system in terms of airport operations. A potential safety hazard is created by wake vortices of large aircraft, for which the only current practical solution seems to be to increase longitudinal separation of aircraft. An increase in separation standards means a decrease in airport capacity.

The data in Figure 4.8 identify thunderstorms as the largest single category causing air

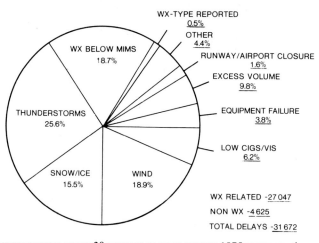

FIG. 4.8. CAUSES OF AIR TRAFFIC DELAYS OF 30 MINUTES OR MORE DURING 1975 IN THE USA (WX, WEATHER; WX BELOW MIMS, WEATHER CONDITIONS BELOW MINIMUM STANDARDS; LOW CIGS/VIS, LOW CLOUD CEILINGS, POOR VISIBILITY, (After Bromley, 1977; reproduced from the *Bulletin of the American Meteorological Society*, p. 1156, with the permission of the American Meteorological Society).

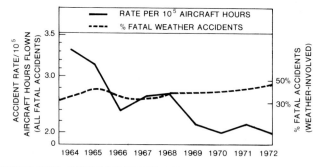

FIG. 4.9. GENERAL AVIATION FATAL ACCIDENTS IN THE USA IN WHICH WEATHER WAS INVOLVED (After Bromley, 1977; reproduced from the *Bulletin of the American Meteorological Society*, p. 1157, with the permission of the American Meteorological Society).

traffic delays, and they also seem to play a major role in weather-related accidents. Turbulence, hail and wind shear within storm clouds have undoubtedly damaged many aircraft, with wind shear recognized as one of the major aviation hazards in the airport environment (Hardesty et al., 1977). A major source of dangerous wind shear is the thunderstorm gust front, a cold air outflow from the thunderstorm downdraft (Bedard et al., 1977; Fujita and Byers, 1977). The gust front is particularly hazardous not only because of the large surface wind shears associated with it, but also because of its very localized nature. The strongest downdrafts, those most likely to be hazardous to aircraft during takeoffs and landings, have been termed downbursts by Fujita and Byers (1977) and identified as vertical currents exceeding a downward speed of $3·6$ m.s^{-1} at 91 m above the surface. The areal extent of a downburst is 800 m or larger in diameter, since an aircraft would very rapidly fly through any smaller area. The violent cloud systems that produce downburst cells can be identified in the form of forward extensions of radar echoes, designated spearhead echoes, moving with unusual speed.

A study by Fujita and Caracena (1977) of three weather-related aircraft accidents shows that all were associated with downbursts encountered by aircraft either descending close to landing or climbing soon after takeoff. All lost altitude while experiencing strong wind shear inside downburst cells. The vertical current of a downburst spreads outwards as it approaches the surface, so that a landing aircraft will first encounter a headwind, then the core of the downburst and finally a tailwind (Fig. 4.10). A similar sequence is experienced by an aircraft immediately after takeoff.

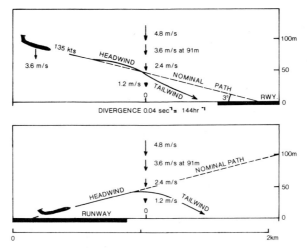

FIG. 4.10. SCHEMATIC DIAGRAMS OF FLIGHT PATHS UNDER THE INFLUENCE OF A DOWNBURST CELL
(After Fujita and Caracena, 1977; reproduced from the *Bulletin of the American Meteorological Society*, p. 1168, with the permission of the American Meteorological Society).

The lessons from such events point clearly to the need for adequate warnings to be provided. In view of the rapid development of downburst cells, communications lags of even a few minutes can prove disastrous. Experiments are being conducted with hybrid acoustic-microwave Doppler radar systems for wind shear measurement and with arrays of pressure sensors for thunderstorm gust front detection (Bedard, et al, 1977; Hardesty et al., 1977) but in the immediate future the hazards still loom large.

Until the last few years the potential hazard of clear air turbulence (CAT) received more attention than the problem of thunderstorm turbulence. This was particularly the case during the 1960s when the large scale use of supersonic transports seemed imminent, although CAT is more likely to be encountered at the lower flight levels of the large subsonic jets (Roach, 1970). The most favourable regions in the atmosphere for formation of CAT are those of rapid vertical change of the horizontal wind, usually found above and below jet streams. The most dangerous form of CAT seems to occur when a jet stream passes over mountainous regions.

The effective, safe and economic use of aircraft and airports relies in large measure on the adequate prediction of atmospheric hazards. Apart from the obvious dangers to life and property and the associated financial losses, knowledge of potential hazards such as icing, strong winds and poor visibility is important to permit the most efficient utilization of resources. If an aircraft has to be diverted from its original destination because of bad weather, considerable expense may be incurred in extra fuel costs, accommodation and transport costs for passengers stranded at the wrong location, and in terms of enforced redeployment of available aircraft. Even a delay because of bad weather can incur additional costs as well as producing poor consumer reaction, with possible future spin-off in lost revenue to the airline in question because of adverse publicity.

From a practical point of view the effective use of ships and ports is just as dependent on the weather. It has been estimated that the loss of time incurred by merchant vessels due to strong winds and rough seas, and by other ships due to various meteorological and oceanographic causes at seaports and their approaches, exceeds 15 per cent of the total budget of the operational time of the fleet (Anon, 1976a).

There is, however, considerable potential for greatly reducing damage and the losses of money and time caused by unfavourable weather and ocean conditions. This potential lies mainly in improvement of meteorological services and in proper use of available weather information. In 1967, weather services in the USA provided weather routing for nearly 4000 ships, resulting in a saving of about US$ 13 million. Present weather routeing caters for about 10 000 ships each year, with savings in time, reduction of accidents at sea and losses of cargo, and reduction of damage to ships.

SUGGESTIONS FOR FURTHER READING

BECKWITH, W. B., 1976, 'The early days of airline meteorology.' *Bull. Amer. Met. Soc.*, 57, pp. 1327–9.

BOOTSMA, A., 1976, 'Estimating minimum temperature and climatological freeze risk in hilly terrain.' *Agric. Met.*, 16, pp. 425–43.

BRADDOCK, R. D., 1970, 'On meteorological navigation.' *J. Appl. Met.*, 9, pp. 149–53.

BROMLEY, E, 1975, 'Aviation weather forecasts in tomorrow's flight service system.' *Bull. Amer. Met. Soc.*, 56, pp. 372–4.

BROWN, L. R., 1975, 'The World food prospect.' *Science*, 190, pp. 1053–9.

CANHAM, H. J. S., 1966, 'Economic aspects of weather routing.' *Marine Observer*, 36, pp. 195–9.

CHASE, P. H., 1970, 'Fifty years of weather insurance.' *Weather*, 25, pp. 294–8.

COLLIS, R. T. H., 1975, 'Weather and world food.' *Bull. Amer. Met. Soc.*, 56, pp. 1078–80.

COUGHLAN, M. J. and LEE, D. M., 1978, 'The assessment of drought risk in northern Australia.' Paper presented to ANU/NARU Seminar on Natural Hazards Management in Northern Australia, Darwin, 11–14 September 1978.

CUMMING, J. N., 1966, 'The effects of the 1965–66 drought on sheep numbers, and the expected rate of recovery.' *Quarterly Rev. Ag. Econs.*, 19, pp. 169–76.

DONALDSON, G. F., 1968, 'Allowing for weather risk in assessing harvest machinery capacity.' *Amer. J. Agric. Econ.*, 50, pp. 24–40.

DONALDSON, R. J. and WEXLER, R., 1969, 'Flight hazards in thunderstorms determined by Doppler velocity variance.' *J. Appl. Met.*, 8, pp. 128–33.

FINKELSTEIN, J., 1971, *The 1969–70 droughts in New Zealand*. Technical Note 204, (New Zealand Meteorological Service, Wellington).

GENERAL AVIATION SAFETY COMMITTEE, 1979, 'Aircraft accidents related to weather.' *Weather*, 34, pp. 269–74.

GREEN, F. H. W., 1978, 'Water levels in England in recent years.' *Weather*, 33, pp. 97–101.

GUSTAFSSON, Y., 1977, 'Variations in rainfall as a natural constraint on agriculture.' *Ambio*, 6, pp. 34–5.

HILL, H. W., 1971, *Some synoptic aspects of drought in New Zealand in the summer 1967–70*. Techanical Note 194, (New Zealand Meteorological Service, Wellington).

IDSO, S. B., 1974, 'Climatic effects of increased industrial activity upon the world's established agro-ecosystems.' *Agro-Ecosystems*, 1, pp. 7–17.

KATZ, R. W., 1977, 'Assessing the impact of climatic change on food production.' *Climatic Change*, 1, pp. 85–96.

LINDQUIST, J. A., 1977, 'Automation—some potential applications to aviation weather.' *Bull. Amer. Met. Soc.*, 58, pp. 1161–3.

McGUIRE, K., 1962, 'Economic effects of drought on 12 properties in the W.A. pastoral zone.' *Quarterly Rev. Ag. Econs.*, 15, pp. 87–94.

MARINE DIVISION, METEOROLOGICAL OFFICE, 1977, 'The development of ship routeing and its modern application at the Meteorological Office, Bracknell.' *Marine Observer*, 47, pp. 23–31.

MOORER, T. H., 1966, 'Importance of weather to the modern seafarer.' *Bull. Amer. Met. Soc.*, 47, pp. 976–9.

PARRY, M. L., 1975, 'Secular climatic change and marginal agriculture.' *Trans. Inst. Brit. Geogrs.*, 64, pp. 1–13.

PULLIN, J., 1976, 'Counting the cost of the dry summer of '75.' *Surveyor*, 147, pp. 12–14.

SCHERHAG, R. WARNECKE, G. and WEHRY, W., 1967, 'Meteorological parameters affecting supersonic transport operations.' *J. Navigation*, 20, pp. 53–63.

SPILLANE, K. T., 1967, 'Clear air turbulence and supersonic transport.' *Nature*, 214, pp. 237–9.

THOMPSON, L. M., 1975, 'Weather variability, climatic change, and grain production,' *Science*, 188, pp. 535–41.

5
Weather Day-by-day

The emphasis in this chapter is on the ways in which day-to-day weather events and their accumulation impinge upon a wide range of human activities. Again, there is no intention to produce an exhaustive catalogue of impacts, but rather to provide some illustrations of the range and hopefully to stimulate greater thought about the importance of the atmospheric resource.

5.1 AGRICULTURE

It is still generally accepted that weather is the most important variable in agricultural production, even though technological advances and improvements in forecasting have made possible minor adjustments in planting and harvesting schedules (Sewell et al., 1968). Landsberg (1968) pointed out that the weather variable enters in two ways into the problem of adequacy of food supplies. One is through weather hazards to crop plants, such as frosts, hail, drought, or pest-provoking conditions. The other is through land use which is wasteful of climatic potential, defined as attempts to raise certain crops in environments for which they are not particularly suited. Properly adapted crops are much more efficient land and energy users than forced crops.

The prominent role of weather in the basic inputs to the crop production system is illustrated in Figure 5.1. Weather elements also appear as vital components in the control complex of the system, with less significant contributions at other points in the system. In a survey of climate, weather and human food systems, Duckham (1974) lists climate, soils, terrain, natural vegetation, pests and diseases as the major biophysical, and, in part, social and economic constraints on human food systems. Figure 5.2 outlines the relationships between climate, weather uncertainty, indirect climatic influences on soils, natural vegetation and pests and diseases, and human food systems. The various constraints on human food systems limit the species which can be produced and can be used for human food. They also limit the possible human food chains and farming systems and through them the amount, efficiency and stability of human dietary output. Constraints attributable to weather and climate can be mitigated or aggravated through the use, or mis-use of available resources and of inputs such as irrigation, mechanization and fertilisers (Duckham, 1974).

The main climatic elements upon which crop growth depends are solar radiation, temperature and moisture. The maximum amount of plant tissue that can be photosynthes-

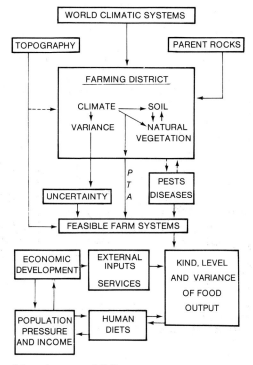

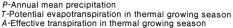

P-Annual mean precipitation
T-Potential evapotranspiration in thermal growing season
A-Effective transpiration in thermal growing season

FIG. 5.1. A SCHEME FOR SYSTEMS ANALYSIS OF LAND USE IN AGRICULTURE
(After Landsberg, 1968; reproduced with the permission of Elsevier Scientific Publishing Co. and Professor H. Landsberg).

ized within a crop depends on the availability of suitable radiation, assuming unlimited carbon dioxide, water and soil nutrients (Fig. 5.3). The growth of many plants responds to day length as well as total available solar radiation. Plants like rice which grow in low latitudes where the day length varies only slightly may be sensitive to day length differences of only a few minutes, while others are indifferent to day length (Linacre and Hobbs, 1977).

Temperature effects on crop growth can operate in several ways. Many crops, such as coffee, bananas and sugar cane are particularly sensitive to frosts, although apples need sufficiently low temperatures to initiate the development of flowers. Most crops cannot be grown successfully unless temperature exceeds some critical threshold value. Coconuts and pineapples thrive only when temperatures are always above 21 °C, and maize requires daily mean temperatures above this for at least part of the growing season. Citrus fruit, cotton, sugar cane and rice grow little at temperatures below 15 °C, many vegetables need at least 8 °C and the threshold for wheat is about 3 °C. The length of the growing season is sometimes taken as the time between the last frost of spring and the first frost of autumn, sometimes as the period for which daily mean temperatures exceed some arbitrary value. If this value is taken as 6 °C and the period calculated as the time exceeded in 80 per cent of

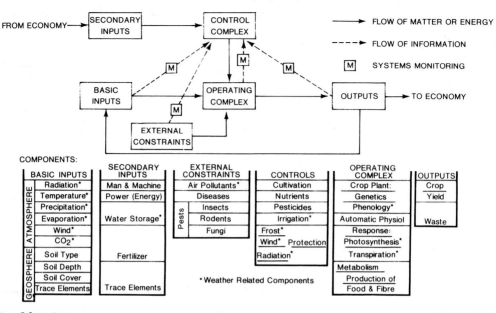

FIG. 5.2. CLIMATE AND FOOD. INTER-RELATIONS BETWEEN CLIMATE, CLIMATE-ASSOCIATED FACTORS, FEASIBLE FARMING SYSTEMS, ECONOMIC DEVELOPMENT, POPULATION, AND HUMAN DIETS
(After Duckham, 1974; reproduced with the permission of the Royal Meteorological Society).

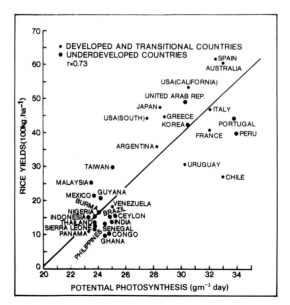

FIG. 5.3. RELATIONSHIP BETWEEN YIELDS OF RICE AND THE POSSIBLE RATE OF PHOTOSYNTHESIS
(After Chang, 1970; reproduced by permission from the *Annals of the Association of American Geographers*, Vol. 60, 1970, Jen-hu Chang).

years, it is found that in England the period is reduced by two to six days for each 10 m of extra elevation. Increasing elevation also increases the time needed for a crop to mature, so that combination of increasing time needed and decreasing time available leads to a ceiling height above which the crop cannot be grown. This ceiling is higher at lower latitudes.

Soil temperature is more important than that of the air for most crops, like potatoes, and for seed germination. Potatoes need at least 8 °C, no more than 28 °C and preferably about 18 °C. Cotton seeds need the soil to be at least at 10 °C for germination, but ideally require soil temperatures below 18 °C since bollworms emerge from the soil at that temperature (Linacre and Hobbs, 1977).

The concept of degree-days allows estimation of the suitability of a crop for a particular climate assuming ample water availability (e.g. Robertson and Coulter, 1973). If the threshold temperature for a crop to grow is 6 °C, a day of 16 °C would count as 10 degree-days. The total number of degree-days over a specified period is called the accumulated temperature and gives an approximate index for comparing the thermal needs of plants, as well as for indicating the likely suitability of a location for a specific crop.

Crop growth depends on ample soil moisture as well as on appropriate temperature conditions. The state of soil moisture governing crop growth is itself governed by rainfall and the evaporation rate. The evaporation rate is related to temperature, wind speed and atmospheric humidity, and the usefulness of rainfall to a crop depends on the stage of development. In general, steady and moderately moist conditions are more favourable to crop growth than an erratic alternation of very wet and very dry spells, so comparison of long-term rainfall and evaporation patterns does not give a very good guide to soil moisture availability.

It is not possible here to examine the specific relationships between all crops and weather and climate, so a few examples are taken as illustrations.

The complexity of yield–weather relationships is clearly shown by wheat which is the most widely grown agricultural crop. The optimum temperature for wheat is 25–31 °C, with about 1200 degree-days above 5 °C needed for typical strains. Weather variations account for much of the variability in wheat production from season to season or region to region, but no single weather variable explains all of the yield fluctuations. The vulnerability of wheat to weather conditions over large areas is well illustrated by production records from North America, Australia and the USSR. Serious droughts in North America in 1961–62 and 1967–68 produced average yields of 1326 kg.ha^{-1} and 1597 kg.ha^{-1} compared with expected yields of 1460 kg.ha^{-1} and 1800 kg.ha^{-1} respectively. The total estimated production loss was 11·5 million tonnes. A massive decrease in production in the USSR in 1972 was a result of a harsh winter, followed by heat and drought during the growing season. The 1972–73 production loss was estimated as 16·3 million tonnes. The Australian wheat crop also suffered from drought in 1972–73, with an average yield of only 840 kg.ha^{-1} compared with a record 1507 kg.ha^{-1} in 1966–67 and 1370 kg.ha^{-1} in 1973–74. Australian production losses in 1972–73 were estimated as 2·8 million tonnes. Such weather-based losses are a substantial proportion of the world's average annual export trade in wheat of 43 million tonnes (Robertson, 1974).

A study of weather and wheat production in the USA indicated the need for higher than normal temperatures in early stages of growth and conditions cooler than normal in the later stages (Thompson, 1969). The same study suggested that spring wheat in North and South Dakota also responds favourably to above normal rainfall throughout the year; that hard red winter wheat in Kansas and Oklahoma responds well to normal precipitation in autumn and winter and again in April with cooler than normal temperatures in spring until the crop matures; and that soft red winter wheat in Illinois and Indiana benefits from normal rainfall throughout the year, near normal temperatures in April, and below normal temperatures in May and June.

Yield per hectare in the Canadian prairie wheat region also varies greatly from year to year, mainly because of weather differences. The 1966 wheat crop of 22·5 million tonnes was almost three times as large as the crop of 1961, and it has been estimated that two thirds of this difference were due to weather differences. Climatic differences within the wheat growing region cause considerable variation in yields and in yield–weather relationships among different parts of the region. In the driest part of the region wheat yields are usually limited by lack of moisture resulting from low precipitation and high potential evapotranspiration. With increasing distance from this semi-arid area, increasing precipitation and decreasing potential evapotranspiration lead to better wheat yields. Beyond a certain optimum distance, however, precipitation is higher and more variable, so that occurrences of excessive precipitation are more common. Such excesses depress wheat yields by delaying planting and harvesting, slowing progress towards maturity, increasing susceptibility to disease, hampering weed control operations and leaching out fertility (Williams, 1971–72).

In Australia rainfall is the chief determinant of wheat yield, the crop being grown mostly in regions where annual rainfalls are between 300 and 700 mm. Table 5.1 shows the effects of rainfall on wheat yields in New South Wales over a six year period. Here the optimum rainfall seems to be in the range from 300 to 450 mm per winter (April to September).

A study of yield data for barley, oats, wheat and potatoes in north-east Scotland has shown a steady upward trend from 1963 to 1974, based on a 5-year moving average. While

Table 5.1
THE EFFECT OF TOTAL RAINFALL[a] FOR APRIL TO SEPTEMBER ON THE SUBSEQUENT WHEAT YIELDS, IN NINE STATISTICAL SUB-DIVISIONS OF NEW SOUTH WALES, FOR SIX HARVESTS, 1969–1975
(after Linacre and Hobbs, 1977)[b]

Range of rainfalls (mm)	Number of cases	Yield (t.ha^{-1})		
		minimum	median	maximum
0–100	1	—	0·16	—
100–150	6	0·33	0·86	0·99
150–200	6	0·58	0·82	1·40
200–250	14	0·86	1·29	1·54
250–300	11	0·91	1·31	1·45
300–350	4	1·04	1·19	1·32
350–400	6	1·40	1·48	1·60
400–450	3	1·59	1·71	1·74
450–550	3	1·25	1·27	1·50

[a] The rainfall for a region for the six months in a particular year is an average from about eleven recording stations within the region

[b] derived from information supplied by the Deputy Commonwealth Statistician, Sydney

this trend has been attributed to the use of improved varieties, higher rates of fertilizer application and generally better husbandry, the year-to-year yield variations from the trend can be largely explained by weather effects (Anderson, 1976). Yields of barley, for example, seem to show a strong correlation with the potential soil moisture deficit and accumulated air temperature above 6 °C in June and with soil temperature at 100 mm, in April and May. The overall evidence is that higher yields resulting from better management cannot give the farmer immunity from weather effects, so that unfavourable weather conditions will limit all crops.

An attempt to understand how weather parameters affect rice yields has shown that the crop reacts differently during different stages of development (Huda et al., 1975). Above average weekly rainfall is beneficial during the nursery period, but during the vegetative phase, coinciding with heavy rainfall at the onset of the monsoon, above average totals have an adverse effect. The ripening phase is most susceptible to excess rainfall (Fig. 5.4). Above average maximum and minimum daily temperatures have beneficial effects during the nursery period, but generally adverse effects thereafter, although departures above average maximum temperatures have decreasing effects during the reproductive and ripening stages (Fig. 5.5). Similar effects on yield are produced by departures from the means of maximum and minimum relative humidities (Fig. 5.6).

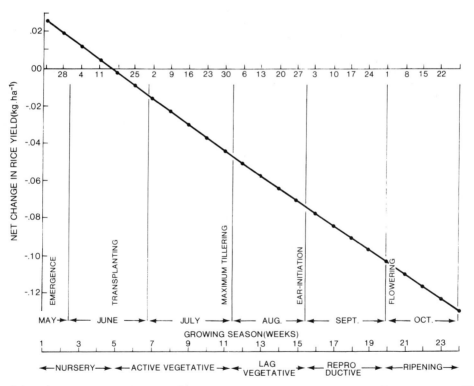

FIG. 5.4. EFFECT ON RICE YIELD OF RAINFALL 1 mm ABOVE AVERAGE WEEKLY TOTAL FOR EACH 7 DAY PERIOD, MAY 21 TO NOVEMBER 4
(After Huda et al., 1975; reproduced with the permission of Elsevier Scientific Publishing Co.).

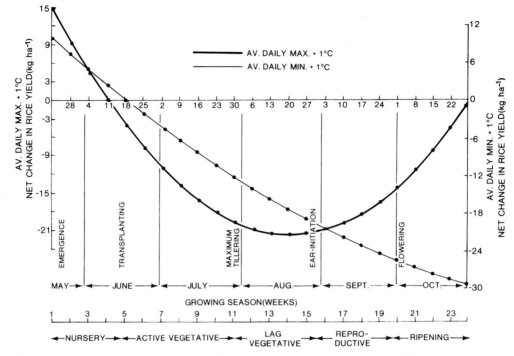

FIG. 5.5 EFFECT ON RICE YIELD OF TEMPERATURES 1 °C ABOVE AVERAGE DAILY MAXIMUM AND 1 °C ABOVE AVERAGE DAILY MINIMUM FOR EACH 7 DAY PERIOD MAY 21 TO NOVEMBER 4
(After Huda et al., 1975; reproduced with the permission of Elsevier Scientific Publishing Co.).

The main meteorological effects on milk yields in England and Wales are the type of spring, the summer rainfall and rainfall during the hay-making season, which determines the quality of winter feed (Smith, 1968). Using such meteorological data combined with milk yield records Smith found it possible to forecast the twelve-month yield at the end of March, with greater accuracy by the end of June. Milk yields at Bellbrook in New South Wales have similarly been found to depend on the production of pasture resulting from rain in autumn and spring. Butterfat production in part of New Zealand in the January to April period over six years varied between 9000 and 35 000 t in response to rainfall differences during the preceding three months. Milk yield in a hot climate is reduced by heat stress or by dry pasture.

There have been many other studies of the effects of weather factors on crop and animal production. For example, an examination of the influence of weather factors on the yield of tea in the Assam valley showed that rainfall up to 180 mm and a rise in mean temperature during the January to March period proved most beneficial to the early crop, leading to an increase in the main crop. Increase in rainfall during this period proved more beneficial when the mean temperature was high rather than low (Sen et al., 1966). Maunder has produced numerous studies of relationships between weather and climate and New Zealand agricultural production. Some examples of the effects of climatic factors on

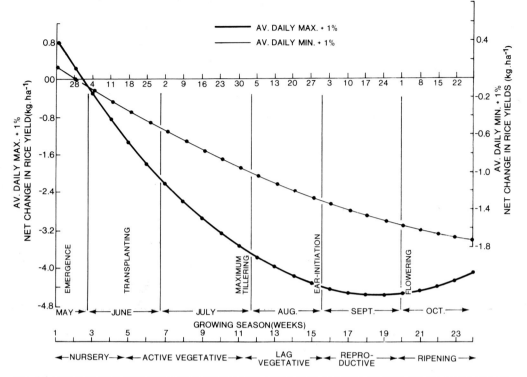

FIG. 5.6. EFFECT ON RICE YIELD OF RELATIVE HUMIDITY 1% ABOVE AVERAGE DAILY MAXIMUM AND 1% ABOVE AVERAGE DAILY MINIMUM FOR EACH 7 DAY PERIOD MAY 21 TO NOVEMBER 4
(After Huda et al., 1975; reproduced with the permission of Elsevier Scientific Publishing Co.).

agricultural production and incomes are shown in Table 5.2. For a country such as New Zealand where perhaps 90 per cent of the foreign exchange comes from the sale of agricultural produce even these relatively small variations are a major economic concern.

Agricultural managers have an important responsibility to determine the extent to which particular operations may or may not be weather sensitive. Applications of meteorological information to management of weather sensitive agricultural processes may include, for example, the use of climatology to adjust planting practices for new genetic characteristics of plants; management of animal production; reduction of storage losses of grain; choice of correct size and type of farm machinery; management of irrigation schemes or of dry-land agriculture; reduction of losses from freezing; mitigation of plant disease impacts; and crop yield forecasting.

Overall, the agriculturalist is probably governed by weather and climate more than most people. Not only does climate largely determine the type of agriculture that is viable, but it also brings risks of uncertainty. In addition weather parameters have an influence on all stages of the agricultural production chain, including not only growth of crops or animals, but also sowing, management, harvesting, transport, and marketing.

Table 5.2
EFFECTS ON PRODUCTION AND AGRICULTURAL INCOMES OF SIGNIFICANT
CLIMATIC FACTORS[a]
(after Maunder, 1968)

Agricultural factor	County	Month/season	Desired Climate[b]	Effect of climatic variations[c]	
				Yield/unit	Value/unit ($NZ)
Wheat	Southland	January	Cloudy[d]	5·3 + 1·1 bushels/acre	$7·2 + 1·4/acre
Oats	Levels	December	Cool	5·3 + 1·6 bushels/acre	$4·4 + 1·4/acre
Barley	Waimea	December	Cool	3·8 + 1·2 bushels/acre	$3·4 + 1·0/acre
Potatoes	Hawkes Bay	December	Sunny	1·14 + 0·42 tons/acre	$88 + 32/acre
Peas	Springs	November	Dry	3·2 + 1·2 bushels/acre	$1·8 + 0·6/acre
Corn	Hobson	November	Cloudy	4·3 + 1·5 bushels/acre	$7·0 + 2·4/acre
Butterfat	Waikato	January	Wet	10 + 2 lb/cow	$2·8 + 0·6/cow
Wool/sheep	Masterton	Previous autumn	Dry	0·3 + 0·1 lb/sheep	$0·14 + 0·04/sheep
Wool/acre	Hawkes Bay	Previous winter	Dry	1·3 + 0·4 lb/acre	$0·6 + 0·2/acre

    [a] significant at the 2·5 per cent level
    [b] relative climate associated with above average production
    [c] one standard deviation above or below average as relevant
    [d] defined as a month with a sunshine duration one standard deviation or more below the average; similar
    definitions used for 'warm', 'sunny', 'wet', 'dry', and 'cool'

5.2 UTILITIES

There are several aspects to the question of relationships between public utilities such as water, electricity, gas and sewerage services and weather parameters. On the one hand there are the relatively obvious impacts of weather on supply of and demand for these services; and on the other hand consideration has to be given to the positive roles of weather elements themselves in providing power sources, particularly from wind and through solar radiation. A further aspect worth examining relates to the possible atmospheric benefits which may accrue through the implementation of new power sources such as natural gas, instead of coal or wood.

Maintenance of adequate and steady water supplies is an important element in the successful functioning of industry, agriculture and domestic households. Urban water supply systems support most industrial productions and the majority of our human resources. An Australian example will be taken here to illustrate the nature of the relationships between climate and water use in urban areas. It seems safe to assume that similar relationships will hold elsewhere.

Urban water use in Australia accounts for only about 10 per cent of the total agricultural, industrial and domestic consumption, but at least 80 per cent of the population and nearly all industrial production depend on it. A study of water use in Australian cities has demonstrated that the number of raindays rather than rainfall per year is the climatic variable most related to total annual per capita water use. Similarly, the number of raindays per month is more significant than monthly rainfall in predicting monthly water use. In cities with hot dry summers mean maximum monthly temperatures and/or monthly pan evaporation are the climatic variables most related to monthly water-use. Garden watering represents 30–50 per cent of mean annual usage in residential areas. On a weekly basis, total pan evaporation is the most significant climatic variable influencing garden use, so there is a clear link with seasonal variations in climate. Mean maximum daily temperature is generally more significant than total weekly rainfall (McMahon and Weeks, 1973).

The relationship between water supply and weather is often rather more complex than this type of analysis may suggest. While increased demands for water are usually related to hot and/or dry weather, failure of supply to meet such demands is not necessarily caused by a basic lack of available water. In many major urban areas, such as Sydney, water held in storage is theoretically sufficient to meet demands for several years, without replenishment. Breakdowns in supply are most likely to relate to overloading of the system, perhaps causing pumps to fail, so that water restrictions have to be imposed. These restrictions are certainly an indirect product of the weather conditions, but contrary to popular belief, are not always caused simply by lack of available water.

The weather may affect power supply in a relatively spectacular and disruptive manner such as occurs during thunderstorms, tornadoes or snowstorms. The more common elements of daily weather such as cloudiness, temperature and wind can also have a bearing upon the amount of electricity consumed and thus be of importance to the authority providing the power.

Electric power consumption is closely related to daily temperatures in distribution areas (Johnson et al., 1969). The relationship between temperature and demand for electrical power is particularly important during the summer months when there are increasing demands for electrically-powered air-conditioning and refrigeration systems. In Perth, in February 1976, for example, heatwave conditions led to electricity consumption exceeding any previous winter record (Gentilli, 1976). Increased demands and overloading of the supply system at such times can lead to the imposition of restrictions on power use and even to breakdowns, all tending to aggravate the existing situation.

The dependence of daily heating requirements upon daily temperature is well known and has been clearly demonstrated in an interesting study by Turner (1968). Data on wintertime emissions of sulphur dioxide from residential and commercial space-heating sources were related to daily temperatures (Fig. 5.7). The hourly rates of fuel use were also related to temperature. The diurnal patterns for days that were coldest at the end of the day differed considerably from patterns for days which were warmest at the end of the day, suggesting a relationship between heating requirements and temperature on a scale smaller than 24 hours. Other factors such as wind speed, solar radiation, user habits and time lag after a temperature change also affect the diurnal variation of space heating. although temperature is probably the most important.

The rising cost of fuel and the increasing concern about the environmental impacts of traditional fuels such as coal and oil, as well as the controversy about nuclear power, have generated growing interest in the use of solar and wind power. Electricity generation using solar power is usually inefficient and expensive, or mechanically complex, so it is more realistic to consider the use of solar energy for heating purposes. A typical family home in Britain uses power at the rate of about 50 kW.day^{-1} in winter and 30 kW.day^{-1} in summer, with the demand for power to provide space and water heating for domestic purposes running at about 40 kW.day^{-1} in winter (Duncan, 1977). Duncan calculated that for a house at Eskdalemuir (55°19′N) 70 m^2 of heat collector at an angle of 35° would be needed to supply 40 kW.day^{-1} for heating in December, with a gross overproduction of heat during summer (Fig. 5.8).

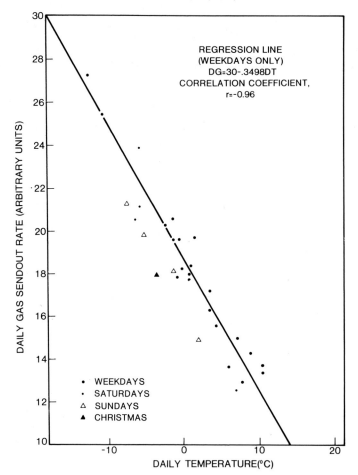

FIG. 5.7. DAILY GAS-SENDOUT RATE AS A FUNCTION OF DAILY TEMPERATURE
(After Turner, 1968; reproduced with the permission of Pergamon Press Ltd.).

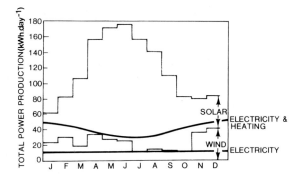

FIG. 5.8. POWER REQUIREMENT OF A TYPICAL HOME, SHOWING ELECTRICITY (10 kWh.day^{-1}) AND ELECTRICITY PLUS HEATING. ALSO SHOWING POWER PROVIDED BY WIND AND SOLAR HEAT COLLECTOR
(After Duncan, 1977; reproduced with the permission of the Royal Meteorological Society).

There are many problems associated with electricity generation by the wind, notably the high capital cost of the installation, the variability of power available, and the difficulty of controlling the frequency and voltage of the electricity generated (Buick et al., 1976). On the other hand, once established, the fuel is free and inexhaustible although limited by the strength of the wind at any particular time.

Location is obviously a critical element in the efficiency of a windmill. Wind speed over urban terrain at 5 m above the ground is about 70 per cent of the speed at 10 m, with the power thus reduced to 34 per cent of the 10 m value. The power reduction is generally not so large over rural terrain. A windmill at Eskdalemuir in an average July at a height of 5 m would need an area of 8·6 m² (radius 1·7 m) to produce 10 kW.day^{-1} at 100 per cent efficiency (Duncan, 1977). In an urban area a windmill of radius 2·2 m would be required. A more realistic efficiency of 25 per cent for electricity generation by windmill would require a doubling of the radius. Clearly, therefore, few sites in Britain would be suitable for wind power generation of electricity on a domestic scale, even ignoring the major problem of storage of the electrical power. In his Eskdalemuir example Duncan calculated that to produce 10 kW.day^{-1} during the minimum wind period in July would need a windmill of 3 m radius mounted at 10 m. Figure 5.8 shows the average monthly power that could be produced by a combination of such a windmill and a 70 m² solar collector. The arrangement certainly seems feasible, although expensive on a domestic scale, and no consideration has been given to factors such as day-to-day or year-to-year fluctuations of wind and solar radiation. Augmentation of heating supplies in Britain in summer appears a reasonable possibility, but year-round self-sufficiency looks unattainable.

Just as increased use of solar and wind power would undoubtedly bring about reduced levels of atmospheric pollution, to the extent that they would reduce the usage of fossil fuels, so also there is evidence that other fuel source changes have similar benefits. Catchpole and Milton (1976) proposed that an increase in the annual duration of bright sunshine in the Canadian prairies in the middle and late 1950s could be attributed to a change in domestic heating fuels from solid fuels to natural gas (Fig. 5.9). The argument is that reduced consumption of solid fuel led to a decline in pollution and thus in cloud formation to such an extent that a significant increase occurred in the duration of bright sunshine. It is interesting to speculate that such a reduction in cloud amount, perhaps brought about by changing fuel consumption patterns, could also lead to more efficient use of solar power because of the increased availability of solar energy.

5.3 RETAILING

Many spheres of business activity are affected in some way by the atmosphere. The overall regional climate may determine the demand for goods such as heaters, air-conditioning equipment and different types of clothing. Areas closely involved with agriculture in particular may experience distinct seasonal cycles in job opportunities and hence in personal incomes and purchasing power. As Smith (1975) mentions, despite what might seem an obvious dependency of industry on the weather and climate, most of the published material relating to commerce concerns retailing or insurance. There is wide recognition that the weather does play an important role in influencing the demand for

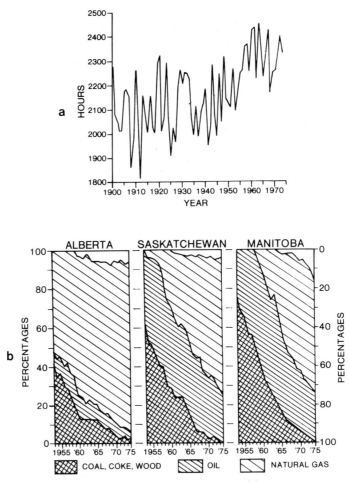

FIG. 5.9. (a) ANNUAL DURATION OF BRIGHT SUNSHINE AT WINNIPEG AND (b) CHANGING PATTERNS OF DOMESTIC FUEL CONSUMPTION IN THE PRAIRIE PROVINCES
(After Catchpole and Milton, 1976; reproduced with the permission of the Royal Meteorological Society).

goods, although many other economic, psychological, sociological and convenience factors also play their parts (Maunder, 1970).

It is possible to envisage several ways in which weather and climate parameters might directly influence consumer behaviour as well as the supply of goods. Perhaps the most obvious impact of weather on consumer behaviour concerns decisions such as whether or not to go shopping, where and how to go and whether the trip has to be curtailed or altered in any way. Inclement weather is likely to reduce the number and length of shopping trips, with consequent effects upon the quantity and type of goods purchased (e.g. Steele, 1951). It might be expected that shoppers will generally be less discriminating in their choice of goods and will only purchase necessities when inclement weather makes shopping unpleasant. On the other hand, it could be argued that since many more shops are now air-conditioned, shoppers might be induced to remain inside stores longer to escape the weather outside. Clearly, the relationships are complex and seem to have been little studied in recent years, which suggests that many interesting and useful investigations remain to be conducted.

One of the most recent studies is that by Maunder (1973) of retail trade–weather associations in the USA. His results indicated that drier than normal conditions appear to be generally associated with above average retail sales in late winter and early autumn; that wetter than normal conditions appear to be generally associated with above average retail sales in late spring and early summer; that colder than normal conditions appear to be generally associated with above average retail sales in the autumn; and that warmer than normal conditions appear to be generally associated with above average retail sales in spring and early summer.

Two earlier studies by Linden (1959a, b) examined the effects of adverse weather on customer traffic in department stores, and looked at temperature-sensitive merchandise lines and the ways in which weather information might be applied to merchandising. Linden (1959a) concluded that on wet days sales fell by about 8 per cent compared with cloudless days. The effect of rain was found to depend on its duration, with reductions exceeding 15 per cent when the rain lasted longer than 6 hours. Although the time of rainfall had some effect on sales, this was most marked when the morning was wet, so that losses on a day that started badly were greater than on a day which became bad after a good start. Rain on Monday was found to have slightly less effect than on other days of the week. On the benefit side, as might be expected, sales of women's raincoats and umbrellas were stimulated by wet weather, much more satisfactorily than by advertising compaigns.

Linden (1959b) reported that the effects of temperature on department store sales were difficult to measure. He found evidence to suggest that sales do fall at times of temperature extremes, to the extent of about 1·5 per cent with temperatures in excess of 29·4 °C. Suburban shoppers seemed to be more affected, with branch store sales falling as much as 5 per cent at times of high temperatures. The major significance of temperature lies with weather-sensitive goods such as air-conditioners, refrigerators, heaters, soft-drinks, ice-creams, beer and outer clothing, which generally show seasonal sales trends. Petty (1963) identified spring and autumn as the critical seasons, warmer than normal springs and cooler than average autumns providing the best retailing conditions.

It is arguable whether manufacturers, wholesalers and retailers make adequate, or any, use of weather information. Linden (1962) and Maunder (1973) have argued for greater and more effective use of weather information to improve economic returns and service to the customer. Knowledge of the weather reflex of a particular product could be extremely useful in improving marketing tactics. Travelling salesmen might be able to coordinate their visits to retail outlets with weather spells likely to favour sales. Distributors, especially those marketing weather-sensitive items such as soft drinks, might try to coordinate regional distributions with long-term weather forecasts, and advertising and promotional activities could be more closely geared to weather forecasts.

There are occasions when hundreds of millions of dollars have been lost or saved by weather prediction. In February 1978, for example, a giant food corporation in New Jersey was able to get 90 per cent of its deliveries completed before commerce was brought almost to a stand-still by a snow-storm, because it acted on the early-warning advice of a private weather forecast service. Another meteorological service in the USA tells its retailer clients when to stock-up with items such as air-conditioners, snow-tyres and umbrellas. This all points to a continuing need for research into the weather sensitivity of particular products.

5.4 CONSTRUCTION INDUSTRY

The associations between weather, climate and construction can be divided into three broad categories. One is the influence of weather and climate on structural and architectural design, already discussed in chapter 3. Another is concerned with the relationships between weather and climate and planning (see chapter 6) and the third, discussed here, is the economic impact of weather on the day-to-day operations of the construction industry.

Enormous sums of money are involved in the construction industry, which employs large numbers of workers, so possible weather-related losses have to be considered in economic terms of decreased profits and intermittent unemployment accompanied by reduced spending power. Russo (1966) estimated that in 1964 the total value of the construction industry in the USA was about US$ 88 billion, representing more than 10 per cent of the gross national product. An estimated 45 per cent of this total was spent in potentially weather-sensitive areas of construction on outdoors work or using perishable materials (Table 5.3). Although these estimates are now very much out of date they nevertheless serve as indicators of the dimensions of the problems involved. It was also estimated that the USA construction industry could save between US$ 0·5 and US$ 1·0 billion annually by appropriate use of potentially available weather information. This represented about 10 to 17 per cent of the estimated weather-caused losses (Russo, 1966) or between a 50 and 100 per cent increase in profits (Beebe, 1967).

In Britain more than 1·5 million were employed in the construction industry in 1966, producing a total output valued at £2900 million (Broome, 1966; Mason, 1966). The value of time lost was about 3·5 per cent of production, equivalent to the loss of ten working days. In the period from 1963 to 1969 the estimated average winter production

Table 5.3
DISTRIBUTION OF TOTAL ANNUAL CONSTRUCTION VOLUME IN THE USA AND THE
PROPORTION CONSIDERED POTENTIALLY WEATHER SENSITIVE
(US$ billions) (after Russo, 1966)

Construction category	Annual volume	Potentially weather sensitive				Total sensitive (per cent of annual volume)
		Perishable material	On-site wages	Equipment	Overhead and profits	
Residential	17·2	0·960	1·624	0·073	2·141	4·8 (27·9)
General building	29·7	1·928	4·079	0·222	2·670	8·9 (30·0)
Highways	6·6	1·666	1·633	0·773	0·727	4·8 (72·7)
Heavy and specialized	12·5	1·875	3·125	2·500	2·500	10·0 (80·0)
Repair and maintenance	22·0	2·674	3·996	1·386	3·143	11·2 (50·9)
Total (rounded)	88·0	9·1	14·4	5·0	11·2	39·7 (45·1)

loss because of seasonal conditions was nearly £80 million. In addition, an average of 30 000 to 50 000 men were laid off every winter.

Precipitation, temperature extremes and high windspeeds are the major causes of disruption to the construction industry, either singly or in various combinations (Table 5.4). The critical rainfall limit for most outdoor activities seems to be a rate of about

Table 5.4
EFFECTS OF WEATHER ON CONSTRUCTION OPERATIONS
(after Winter Building Advisory Committee, 1971; reproduced with the permission of the Controller of Her Majesty's Stationery Office)

Phenomenon	In conjunction with	Effect
Rain		1. Affects site access and movement 2. Spoils newly finished surfaces 3. Delays drying out of buildings 4. Damages excavations 5. Delays concreting, bricklaying, and all external trades 6. Damages unprotected materials 7. Causes discomfort to personnel 8. Increases site hazards
	High wind	1. Increases rain penetration 2. Reduces protection offered by horizontal covers 3. Increases site hazards
High wind		1. Makes steel erection, roofing, wall sheeting, scaffolding, and similar operations hazardous 2. Limits or prevents operation of tall cranes and cradles, etc. 3. Damages untied walls, partially fixed cladding, and incomplete structures 4. Scatters loose materials and components 5. Endangers temporary enclosures
Low and subzero temperatures		1. Damages mortar, concrete, brickwork, etc. 2. Slows or stops development of concrete strength 3. Freezes ground and prevents subsequent work in contact with it, e.g. concreting 4. Slows down excavation 5. Delays painting, plastering, etc. 6. Causes delay or failure in starting of mechanical plant 7. Freezes unlagged water pipes and may affect other services 8. Freezes material stockpiles 9. Disrupts supplies of materials 10. Increases transportation hazards 11. Creates discomfort and danger for site personnel 12. Deposits frost film on formwork, steel reinforcement, and partially completed structures
	High wind	Increases probability of freezing and aggravates effects of 1–12 above.
Snow		1. Impedes movement of labour, plant, and material 2. Blankets externally stored materials 3. Increases hazards and discomfort for personnel 4. Impedes all external operations 5. Creates additional weight on horizontal surfaces
	High wind	Causes drifting which may disrupt external communications

0·5 mm.h⁻¹, affecting operations from initial site surveys to final exterior painting (Smith, 1975). Low or freezing temperatures also create problems. Green concrete will increase in strength to as low as -11 °C, but frost damage is a danger below about -2 °C and temperatures below 10 °C reduce the rate of hardening. The combination of precipitation and low temperature producing snow increases construction problems. As Broome (1966) emphasized, dealing with a frozen hole is one thing, but the same hole full of snow is another matter altogether. Steel reinforcing rods, incorporated in most concrete structures, have the ability to conduct heat much more readily than snow so that attempts to melt the snow often result in an impenetrable slab of steel-reinforced ice. Wind presents obvious dangers for working on tall structures or when using tall and slender cranes.

Most of the work on relationships between weather and the construction industry has been concerned with the industry in general with relatively little attention to the road building sector. A study of the effects of climatic variables on the highway construction industry in the USA by Maunder et al. (1971a, b) concentrated on the influence of weather on working conditions during the main construction months. They combined daily engineering records from two road construction projects with soil moisture and precipitation measurements to develop a model capable of describing conditions suitable for road building activities. The model was applied to long-term rainfall records to calculate road construction conditions over a 50-year period. Assuming a full working day to be 8 hours, they found that, on average, 70–80 per cent of the total possible time could have been worked between 1918 and 1967, with the monthly average during the construction season varying from 69 per cent in April and May to 82 per cent in July. Since knowledge of working conditions is important for planning all phases of road construction, Maunder et al. (1971b) converted the daily series of working conditions into an index expressed in hours of worktime. They found, for example, that a road building job requiring 1 000 hours of work is most likely to be completed with the least delay if it is started in late June or early July, but is likely to experience the greatest delay if started in late February or early March.

Weather conditions, forecast as well as actual, are clearly of considerable value to the construction industry. Even as early as the tender stage, when costs are being estimated, it is useful to know the likely weather interference over perhaps the next 12 months. While the job is under way, accurate daily, perhaps even 3-hourly forecasts, would greatly assist efficient completion. The very localized scale at which forecasts would be required complicates the forecasting problem (see chapter 7) and emphasizes the possible role for the consulting meteorologist (see chapter 8).

5.5 LEISURE AND RECREATION

The influence of weather in determining suitability of areas for outdoor leisure and recreation is often assumed as self-evident and therefore to require no elaboration. However, little is known in quantitative terms about the effects of weather conditions on recreation and tourism. In particular, the critical economic impact on commercial prospects for the leisure industries is not well documented. Maunder (1973) has pointed to the need to determine the value of the weather to different activities, but there seems to have been no concerted effort to study the value to the tourist industry. Perry (1972) has suggested that the rapidly increasing demand for weather information from forecasting services is partly the result of growth in leisure time in an affluent society.

Weather and climate can be regarded as part of a regional resource base with significance influencing the functional value of other recreation resources by enhancing or detracting from the intrinsic appeal of a recreation site. Over a period of time, even tourists became aware of, or at least convinced of, apparent climatic deterioration at a resort area and take their custom elsewhere. On the other hand, day-to-day weather conditions can be expected to provoke short-term reactions from impulse visitors. Impulse travel is highly weather sensitive to the extent that there can be marked seasonal variations in the tourist industry, as illustrated for Australia by Tucker (1965).

Weather can have a pronounced seasonal effect on commercial activities associated with tourism. Patronage and economic rewards are often functions of atmospheric conditions and of their occurrence in a predictable manner. Leisure industries are geared to cater for an expected seasonal influx of visitors, from local to international scales of investment. Much of the remarkable growth which has been sustained at tourist resorts in many parts of the world can be traced to the assumption of a pleasant, reliable and appropriate weather pattern. A significant portion of the budget for the tourist industry appears to be allocated to the 'image-makers' for creating the 'desirable' weather picture. Advertising for tourism, and immigration, in much of Australia, for example, often stresses sun, surf and sand, perhaps fostering the impression that these are nationwide, year-round attributes (Pigram and Hobbs, 1975). Expectations fostered by advertising can influence subsequent judgement of actual experiences (Perry, 1972).

Outdoor recreation involves discretionary use of time, money and effort in a setting exposed to the elements. The process of choice and decision making is a vital consideration, and conditions of weather and climate must have a strong bearing on it. However, as Maunder (1970) has pointed out, little is known about the extent to which any particular tourist region is affected by weather in that region, by the weather in another tourist region, or by the weather in the tourists' areas of origin.

During the planning and anticipation phase, the potential recreationist may be influenced by weather conditions at possible recreation sites to the extent of his knowledge or perception of those conditions, including his reliance on weather forecasts (Adams, 1973), and the type of activity planned. The time scale involved is important, as much more is at stake in a decision regarding an annual vacation than a beach day-trip at the weekend. When a major trip is being planned, individual experience or 'weather memory' may be influential, so that past weather experiences and information sources may impinge upon the decision. Perry (1972) suggests that the decision whether or not to return to a particular area seems little influenced by weather experience, but the extent to which past experience produces a change of holiday orientation in future is not known. Actual weather conditions prevailing when the alternatives are being considered might also influence decisions.

Weather conditions en route and at the site affect recreation behaviour and could well influence future patterns in the recollection phase, as to route taken, mode of travel, type and duration of activities, and the site itself. Many of these decision variables will be consciously related to the range of services, facilities, shelter and alternative amusements available should weather conditions deteriorate. Weather tolerance levels are, however, subjective, as evidenced by controversy and argument when sporting events are cancelled or some participants fail to appear.

In some parts of the world, actual or potential bodily discomfort may limit tourist potential (Terjung, 1968). Yet little information is available as to when and how particular weather variables affect recreation and tourism. The influence of factors such as temperature, sunshine, wind, and rain obviously differs between activities and sites. Presumably it also varies between individuals and groups of recreationists. Gaffney (1976) identified the main meteorological parameters affecting the physical well-being of humans as temperature, humidity and air movement, with the main composite variables determining the feasibility of outdoor activities being physical comfort and the prevailing weather. Using comfort assessments based on temperature and humidity and a weather classification based on rain and cloud, Gaffney derived compound meteorological ratings to indicate suitability for specified activities (landscape touring, vigorous activity, swimming) for selected locations in Australia (Table 5.5).

This type of analysis is a step towards the identification of critical threshold levels, beyond which participation and patronage are affected, positively or negatively. Paul (1972) analysed the level of daily participation in eight different outdoor pursuits for three climatically diverse areas of Canada and found that the effect of weather varied markedly according to type of recreation available. Swimming and beach use activities were mainly dependent on daily maximum temperatures and total sunshine hours, whereas attendance at multi-activity parks, picnicking and pleasure driving showed only slight relationships with the selected weather parameters. Where relationships are apparent they are not necessarily simple. Beach usage, for example, can be observed to be limited by temperatures being too high as well as not high enough. Smith (1975) suggests that models similar to that of Paul could be developed to the stage where daily attendances at outdoor recreation facilities, associated traffic flows, and demands for services might be predicted.

The effects of weather on sport are generally no better documented than those on tourism and recreation. Thornes (1977a) has attempted to classify the wide range of interrelationships between weather and sport (Fig. 5.10). It is difficult to think of a meteorological variable that does not have some effect on some sport. Temperature, precipitation and wind are obviously important for many sports, but so also are factors like visibility, humidity and sunshine, not to mention the effects they have on playing surfaces. Thornes makes a useful distinction between three broad groups of sports: specialized weather sports, weather interference sports and weather advantage sports.

Specialized weather sports such as sailing, gliding, hot-air ballooning (Samuel, 1972), parachuting (d'Allenger, 1970) and skiing require certain weather conditions in order to take place at all. In sailing, shifts in wind direction and speed variations can win or lose races. A knowledge of wind behaviour in general, and at particular localities, can be advantageous (Watts, 1967). For the 1968 Olympic Games at Acapulco, the British yachting team employed a full-time meteorologist for advice on local winds, and before the Games they consulted climatologists for general information on likely weather conditions (Houghton, 1969; Watts, 1968).

Gliders (both sail-planes and hang-gliders) require assistance from the atmosphere in the form of uplift. Strong thermals present ideal conditions for conventional gliding and an

Table 5.5

METEOROLOGICAL RATINGS[a] FOR SELECTED OUTDOOR ACTIVITIES AT SELECTED LOCATIONS IN AUSTRALIA

(after Gaffney, 1976)

Location	Activities											
	Beaching or Swimming				Vigorous activity				Landscape touring			
	Jan.	April	July	Oct.	Jan.	April	July	Oct.	Jan.	April	July	Oct.
Adelaide	77	57	33	56	53	77	75	77	59	77	65	77
Alice Springs	85	77	63	81	41	61	94	77	47	70	89	67
Brisbane	70	72	60	71	43	53	93	53	46	61	85	70
Broome	77	89	79	95	32	39	73	41	27	41	83	48
Cairns	71	70	67	85	23	27	73	43	19	29	85	51
Canberra	69	56	30	53	59	83	68	79	65	75	51	77
Carnarvon	86	79	62	77	39	51	90	65	51	62	85	79
Ceduna	76	63	47	65	59	75	84	75	63	75	78	79
Charleville	81	75	61	77	38	59	91	59	41	69	85	71
Darwin	64	81	83	85	21	36	63	38	17	27	67	36
Hobart	61	49	28	44	81	80	67	80	77	73	47	63
Kalgoorlie	85	68	47	68	45	73	89	71	53	73	80	77
Launceston	60	52	27	45	81	80	65	79	77	69	43	65
Marble Bar	79	89	67	93	37	39	97	45	33	45	95	49
Melbourne	67	53	29	49	67	75	73	75	67	73	56	71
Mildura	80	66	43	63	49	73	85	74	55	81	82	75
Perth	79	65	43	61	51	67	72	75	61	71	64	78
Sydney	70	61	53	59	55	71	86	76	58	71	75	75
Townsville	73	81	72	84	27	39	80	49	23	45	89	57

[a] Meteorological rating = $\dfrac{(\text{comfort index} + \text{weather index})}{2}$ %; >70 ideal, 60–70 marginal, <60 sub-marginal, where, for example, for landscape touring,

comfort index = $0.5 F_{H(40.0/50.4)} + F_{H(25.5/40.4)} + 0.75 F_{H(25.5/25.4)} + 0.25 F_{H(20.5/25.4)}$; where, in turn, for example,

$F_{H(40.5/50.4)}$ = percentage frequency of occasions with humidex (H = dry bulb temp. °C + wet bulb temp. °C) in range from 40.5 to 50.4 etc.; coefficients 0.5,

0.75, 0.25 represent 50%, 75%, 25% of population satisfied in respective classes of comfort.

experienced pilot can also gain lift from sea-breeze fronts and lee waves. Gliding requires specific weather information and forecasting, indicating the depth of convection, strength and organization of thermals, and about conditions which might inhibit thermal activity (Wallington, 1968; Wickham, 1966).

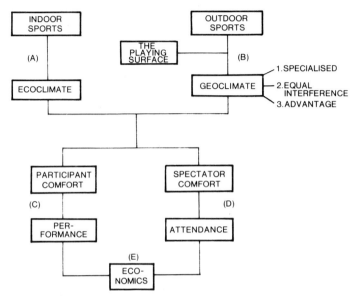

FIG. 5.10. THE EFFECTS OF WEATHER ON SPORT
(After Thornes, 1977a; reproduced with the permission of the Royal Meteorological Society).

Attempts to organize skiing in Scotland in the 1930s coincided with a run of poor snow seasons, and the attempts failed. The present development in the Scottish skiing industry has in part been helped by better recent snow seasons than earlier in the century. As well as determining the amount of snowfall, weather conditions affect the length and condition of the ski runs and the operation of lifts and tows. Chairlifts have to close when the cross-wind velocity exceeds about 15 m.s^{-1}. Perry (1971a) has investigated in some detail the climatic influences on the Scottish skiing industry, concluding that the number and type of climatic problems, coupled with competition from more favoured Alpine resorts, should discourage major capital investment.

Sports such as soccer, hockey, rugby, bowls and lawn tennis fall into the weather interference category. These are ideally suited to 'weatherless' days which should be warm, dry, bright but overcast, with little or no wind, excellent visibility, not too humid and with a firm ground surface (Thornes, 1977a). The weather is an equal interference factor in this case, with all competitors generally having to adjust to the same weather conditions and with matches played in two halves to compensate for factors such as wind direction.

Weather advantage sports include those of unequal interference, such as golf tournaments spread over several hours, with individual competitors encountering different weather conditions favouring some but hindering others. Surfing competitions also fall into this category. Both surfboard and life-saving events usually involve heats at different times. For surfboard events wind conditions are particularly important, with light, offshore winds promoting optimum wave shape and smoothness to help the competitor, but with strong or onshore winds severely affecting wave quality and thus competitor performance.

In some sports, such as cricket, the competing teams are performing different functions at the same time. Conditions assisting the fielding side, such as a rain-affected pitch or a humid atmosphere, are a distinct disadvantage to the batting side, yet the situation may be reversed under different weather conditions.

The state of the playing surface is critical for many outdoor sports and is determined largely by the weather. Grass surfaces are very weather-sensitive, not only to precipitation but also to the effects of frost. Some grounds have been underlain by electric cables for heating to prevent snow and frost cover, and artificial playing surfaces are increasing in number.

Weather conditions are very important in determining participant and spectator comfort. In hot, humid conditions participants experience heat stresses which can severely affect performance. Cold conditions cause blood supplies to the hands and feet to be reduced in far greater proportion than to the rest of the body, so that sensitivity and hence performance suffers. Optimum performance occurs within a thermal comfort zone which varies according to the type of activity.

The most direct economic effect of weather on sport is in terms of attendances. The problem here is that in general a thermally comfortable environment for a participant, say around 12 °C, may well be uncomfortable for spectators, because of their different activity rates. Solutions to this problem, like the Houston Astrodome, tend to be very expensive. Nevertheless, it is clear that sport management should be constantly aware of the effect of weather upon its decisions, so that performances, and hence participant and spectator satisfaction, might be improved and enterprises kept on a sounder economic footing.

SUGGESTIONS FOR FURTHER READING

ARAKAWA, H., 1959, 'Hydroelectric power generation and the climate of Japan—a case of engineering meteorology.' *Bull. Amer. Met. Soc.*, 40, pp. 416–22.

BAIER, W., 1973, 'Crop–weather analysis model: review and model development.' *J. Appl. Met.*, 12, pp. 937–47.

BAILEY, J., 1975, 'Bristol's bridges: macroclimatology and design.' In *Processes in Physical and Human Geography. Bristol Essays.* Eds. Peel, R., Chisholm, M. and Haggett, P. (Heinemann Educational Books, London) pp. 197–217.

BONACINA, L. C. W., 1975, 'Phenology.' *Weather*, 30, pp. 58–60.

BOYD, W. J. R., GOODCHILD, N. A., WATERHOUSE, W. K. and SINGH, B. B., 1976, 'An analysis of climatic environments for plant-breeding purposes.' *Aust. J. Agric. Res.*, 27, pp. 19–33.

BROCK, F. V., 1959, 'Engineering meteorology: the effects of weather and climate on concrete pavement.' *Bull. Amer. Met. Soc.*, 40, pp. 512–17.

BROWN, L. R., 1975, 'The world food prospect.' *Science*, 190, pp. 1053–9.

BUNTING, A. H., 1975, 'Time, phenology and the yield of crops.' *Weather*, 30, pp. 312–25.

CHANG, J.-H., 1968, 'Progress in agricultural climatology.' *Prof. Geogr.*, 20, pp. 317–20.

CHASE, P. H., 1970, 'Fifty years of weather insurance.' *Weather*, 25, pp. 294–8.

CRADDOCK, J. M., 1965, 'Domestic fuel consumption and winter temperatures in London.' *Weather*, 20, pp. 257–8.

DALTON, G. E., 1974, 'The effect of weather on the choice and operation of harvesting machinery in the United Kingdom.' *Weather*, 29, pp. 252–60.

DEWILJES, H. G. and ZAAT, J. C. A., 1968, 'Influence of climate upon the number of weather-working hours in combine harvesting in the Netherlands.' *Arch. F. Met. Geophys. u. Bioklim.*, B, 16, pp. 105–14.

DOLL, J. P., 1967, 'An analytical technique for estimating weather indexes from meteorological measurements.' *J. Farm Econs.*, 49, pp. 79–88.

EWALT, R. E., WIERSMA, D. and MILLER, W. L., 1973, 'Operational value of weather information in relation to soil management characteristics.' *Agronomy J.*, 65, pp. 437–9.

GANDER, R. S., 1972, 'The demand for beer as a function of the weather in the British Isles.' In *Weather Forecasting for Agriculture and Industry*. Ed. Taylor, J. A., (David and Charles), pp. 184–94.

GILLOOLY, J. F. and DYER, T. G. F., 1977, 'Structural relationships between corn yield and weather.' *Nature*, 265, pp. 434–5.

GRIBBIN, J., 1976, 'Climatic change and food production.' *Food Policy*, 1, pp. 301–12.

HASHEMI, F. and DECKER, W. L., 1969, 'Using climatic information and weather forecasts for decisions in economising irrigation water.' *Agric. Met.*, 6, pp. 245–57.

HOGG, W. H., 1964, 'Meteorology and agriculture.' *Weather*, 19, pp. 34–43.

— 1964, 'Weather and horticulture.' *Weather*, 19, pp. 234–41.

— 1967, 'Meteorological factors in early crop production.' *Weather*, 22, pp. 84–94 and 115–18.

— 1970, 'Weather, climate and plant disease.' *Met. Mag.*, 99, pp. 317–26.

JOHNSTONE, D. R. and HUNTINGTON, K. A., 1977, 'Weather and crop spraying in northern Nigeria.' *Weather*, 32, pp. 249–57.

MCENTEE, M. A., 1975, 'Climatic variability and agricultural research.' *Farm and Food Res.*, 6, pp. 142–4.

MAUNDER, W. J., 1966, 'Climatic variations and agricultural production in New Zealand.' *N.Z. Geogr.*, 22, pp. 55–69.

— 1966, 'Climatic variations and dairy production in New Zealand: a review.' *N.Z. Sci. Rev.*, 24, pp. 69–73.

— 1967, 'Climatic variations and wool production: a New Zealand review.' *N.Z. Sci. Rev.*, 25, pp. 35–9.

— 1968, 'Effect of significant climatic factors on agricultural production and incomes: a New Zealand example.' *Mon. Weath. Rev.*, 96, pp. 39–46.

— 1968, 'Agroclimatological relationships: a review and a New Zealand contribution.' *Canad. Geogr.*, 12, pp. 73–84.

MUSGRAVE, J. C., 1968, 'Measuring the influence of weather on housing starts.' *Construction Rev.*, 14, pp. 4–7.

NAYA, A., 1967, 'Insects, insecticides and the weather. Part one.' *Weather*, 22, pp. 139–46.

— 1967, 'Insects, insecticides and the weather. Part two.' *Weather*, 22, pp. 211–15.

NEWELL, R. E., TANAKA, M. and MISRA, B., 1976, 'Climate and food workshop: a report.' *Bull. Amer. Met. Soc.*, 57, pp. 192–8.

NICHOLLS, N., 1978, 'Sunspots and Australian wheat yield.' *Search*, 9, p. 319.

OURY, B., 1965, 'Allowing for weather in crop production model building.' *J. Farm Economics*, 47, pp. 270–283.

ROBERTSON, G. W., 1975, *Rice and weather*. Technical Note No. 144, WMO-No. 423, (World Meteorological Organisation, Geneva).

SHAWCROFT, R. W., LEMON, E. R., ALLEN, L. H., STEWART, D. W. and JENSEN, S. E., 1974, 'The soil–plant–atmosphere model and some of its predictions. *Agric. Met.*, 14, pp. 287–307.

SMITH, L. P., 1967, 'Meteorology and the pattern of British grassland farming.' *Agric. Met.*, 4, pp. 321–38.

— 1975, 'Agricultural aspects of international meteorology.' *Agric. Admin.*, 2, pp. 43–8.

SPURR, G., 1974, 'Meteorology and cooling tower operation.' *Atmos. Env.*, 8, pp. 321–4.

STEPHENS, F. B., 1951, 'A method of analysing weather effects on electrical power consumption.' *Bull. Amer. Met. Soc.*, 32, pp. 16–20.

SUOMI, V. E. 1975, 'Atmospheric research for the nation's energy program.' *Bull. Amer. Met. Soc.*, 56, pp. 1060–80.

THOMPSON, L. M., 1970, 'Weather and technology in the production of soybeans in the central United States.' *Agronomy J.*, 62, pp. 232–6.

— 1975, 'World weather patterns and food supply.' *J. Soil and Water Conservation*, 30, pp. 44–7.

THORNES, J. E., 1976, 'Rain starts play.' *Area*, 8, pp. 105–12.

THORNES, J., WOOD, L. and BLACKMORE, R., 1977, 'To salt or not to salt?' *New Scientist*, 73, pp. 326–8.

WRIGHT, P. B., 1969, 'Effects of wind and precipitation on the spread of foot-and-mouth disease.' *Weather*, 24, pp. 204–13.

6
Weather, Climate, and Planning

It has been demonstrated that considerable economic benefits can flow from efficient utilization of climatological data in commerce and agriculture. It has also been shown how weather and climate can affect human comfort and well-being. Combining factors of economics and comfort points to the important role of meteorology in urban planning, and in the siting of industrial complexes and public utilities such as airports and power stations.

Claims are frequently made that insufficient attention is given to the climatology of the low levels of the atmosphere, revealed by temperature, stability and wind profiles or to the effects of perturbations caused by local topography and large structures. There is increasing emphasis on the quality of the atmosphere in which we live, which is largely determined by its characteristics in the first thousand metres above the ground. This has focused attention on the applications of climatology to urban planning, particularly in relation to air and noise pollution, public health and human comfort, and the effects of high-rise structures on the climate at street level and of the whole urban area.

A growing proportion of the world's population lives in urban environments, so that by the end of this century perhaps 60 per cent of all people will live in towns. In many countries around the world, urban dwellers even now account for as much as 90 per cent of national populations. Whenever any of the commonly used construction materials are substituted for forests or fields, the physical and chemical properties of the atmospheric boundary layer are changed, so that elements such as airflow, temperatures, precipitation, humidity and visibility differ in urban as compared with rural areas. Not only is the air above cities polluted with a mixture of solid, liquid and gaseous materials, but temperatures are generally higher than in the surrounding country, precipitation tends to be increased at least in frequency and sometimes in amount, strong ventilating winds are decelerated and light winds are accelerated, and visibility and radiation receipts are lowered by the pollutants.

Chandler (1976) has emphasized the need for consideration of climate as an essential element in the planning process and has attributed the general neglect of climatic considerations partly to the relatively recent growth of the science of urban climatology, and partly to the weak communication links between climatology and planning. Consideration of the effect of climate on planning arises in two main ways, firstly in relation to the initial siting of new settlements or industries, and secondly in the location of industries or new settlements as part of already established areas. In the first case planners are starting to make use of available data and climatological expertise, but the problem is much greater when dealing with sites originally determined by various geographical, historical, political or economic forces. In addition, planning does not end with site selection, and climatic knowledge has obvious applications to the internal design of urban areas, in the decisions about preferred locations of residential, industrial and recreational zones. Climatological data can also be used to aid the planning of size and space of buildings as well as the designs of the buildings themselves.

6.1 URBAN ATMOSPHERIC ENVIRONMENT

The meteorological properties of the air within and immediately above urban areas are markedly changed by the processes of urbanization, creating distinctive urban climates. The planner has to take account not only of the regional scale climatic characteristics of the site but must also be aware of the ways in which man's constructions modify these to produce a variety of local climates. At the same time the planner must be able to recognize the likely impacts of these modified climates on features like air pollution, patterns of air flow, formation of heat islands and alteration of hydrologic regimes. It is convenient to consider such impacts separately, but this, of course, ignores the fact that in reality they interact with great complexity to make the job of the planner far from easy.

Patterns of airflow within and immediately above cities are fundamental to most meteorological processes and hence to temperature, humidity and precipitation patterns and to levels of pollution (Chandler, 1976). The mean annual wind speed within cities is about 20 to 30 per cent less than that observed at airport sites in predominantly surburban or rural areas, although the reduction is most noticeable when regional winds are relatively strong (Table 6.1). At night or when prevailing winds are relatively light, urban wind speeds may be greater than those in rural areas.

Table 6.1
AVERAGE WIND SPEED AT LONDON AIRPORT AND ITS EXCESS ABOVE THAT
IN CENTRAL LONDON (m.s^{-1})
(after Chandler, 1976)

Season	0100 GMT		1300 GMT	
	Mean speed	Excess speed	Mean speed	Excess speed
December–February	2·5	−0·4	3·1	0·4
March–May	2·2	−0·1	3·1	1·2
June–August	2·0	−0·6	2·7	0·7
September–November	2·1	−0·2	2·8	0·6
Year	2·2	−0·3	2·9	0·7

Studies of the ways in which urban wind directions fluctuate have shown that turbulent diffusion in urban areas is significantly larger than in the country, with the pattern of winds being closely controlled by a combination of site topography and the forms of buildings (Ludwig and Dabberdt, 1973). It is difficult to be certain of the actual extent to which air flow patterns are influenced by cities, although studies in St. Louis have shown that there are quite distinct time and space variations in the wind fields in the lowest 2000 m of the atmosphere over the city (Ackermann, 1974a, b).

In some coastal cities such as Sydney the patterns of airflow are complicated by sea breeze effects during the day and drainage of air towards the coast at night. Thus, during the daytime sea breezes tend to scour pollutants from areas near the coast and move them inland, with the reverse happening at night. The resultant effect is frequently a recirculation of atmospheric pollutants over the city rather than their removal.

Urbanization seems to affect precipitation through increases in hygroscopic nuclei, through mechanical turbulence via increased surface roughness, through increased thermal turbulence because of increased surface temperatures, and possibly through the addition of water vapour from combustion sources. It is, however, difficult to prove urban effects on precipitation compared with that in rural areas since sampling is generally poor, precipitation is notoriously variable over very small distances, and urban effects could extend outside urban areas. Measured increases of annual precipitation in and down-wind of urban areas vary but generally seem to be about 5 to 15 per cent greater, with larger increases in winter than in summer. A series of studies by Changnon observed precipitation regimes and patterns in several USA cities, showing an average annual precipitation increase of nearly 7 per cent. There was a marked increase in the number of days with thunderstorms and hail, similar to that observed in many city areas, including London (Atkinson, 1968, 1969).

On the other hand some studies in North America have suggested that urbanization may not produce a marked increase in precipitation. A study in the New York metropolitan area, for example, indicated no effect on daily precipitation patterns. Nevertheless, it does seem likely that convection related to an urban heat island can produce a significant effect on precipitation in some cities such as Washington, DC, where a study by Harnack and Landsberg (1975) showed that the extra increment of heating supplied to a convective cloud by the urban fabric is often the trigger for the occurrence of convective precipitation.

The existence of detectable thermal anomalies of man-made origin in large towns and cities is well documented and has been reported by many researchers. Apart from the possible exception of pollution, more attention has been given to the study of temperatures in urban areas than to any other meteorological element. Oke (1974) has presented a comprehensive listing and summary of most studies to that time and Chandler (1976) provides a useful discussion of some of the most important features arising from work on urban heat islands. Urban-rural temperature differences commonly reach 5 °C, with heat island effects being generally most intense at night, when the temperature differences may reach 11 °C (e.g. Bornstein, 1968; Peterson, 1969; McGrath, 1971; Oke and East, 1971; Eagleman, 1974; Kalma et al., 1974; Oke and Maxwell, 1975).

Measurements in the USA suggest that a ten-fold increase of city population has an average warming effect on the centre of 1 °C. It is also found that the urban heating effect is removed by winds of sufficient strength, ranging from 4·1 m.s^{-1} for a city of 33 000 population to 11·8 m.s^{-1} for a metropolis of 8 million (Oke, 1973, 1976) (Figs. 6.1 and 6.2).

Several factors may be involved in the production of an urban heat island. Man-made heat may be appreciable, especially in the largest industrialized cities. Some rates of man-made energy production are comparable with net radiation fluxes. In central Sydney,

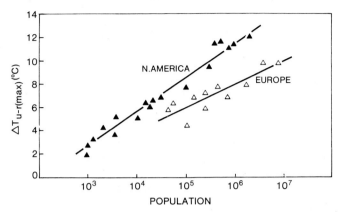

FIG. 6.1. RELATION BETWEEN MAXIMUM OBSERVED HEAT ISLAND INTENSITY AND POPULATION FOR 18 NORTH AMERICAN AND 11 EUROPEAN SETTLEMENTS
(After Oke, 1973; reproduced with the permission of Pergamon Press Ltd.).

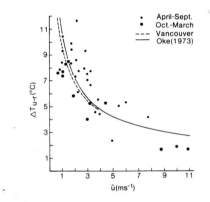

FIG. 6.2. RELATION BETWEEN MEASURED HEAT ISLAND INTENSITY AND THE REGIONAL WIND SPEED FOR VANCOUVER, B.C. DATA FROM 1 TO 3 HOURS AFTER SUNSET ON CLOUDLESS NIGHTS IN ALL SEASONS
(After Oke, 1976; reproduced with the permission of the University of Toronto Press).

in winter, for example, man-made heating can exceed 25 $W.m^{-2}$, equivalent to almost 50 per cent of incoming solar radiation at that time (Kalma, 1974). In summer the figure falls to something like 25 per cent of the incoming solar radiation. A second reason for the heat island effect is the construction materials used in modern cities: concrete, brick, bitumen and rock all readily absorb the daytime heat and freely release it to the atmosphere at night. In addition the relatively rapid drainage of water from a city reduces evaporative cooling. There is also a reduced albedo because of the relative absence or removal of snow, the replacement of vegetation by low-albedo material such as bitumen, and because of the concave surfaces of the city's profile. The intensity of a heat island does seem to relate to the density of buildings as well as to population. Temperatures tend to be higher where buildings are closer together and streets are correspondingly narrow.

The problems posed by air pollution are of obvious importance in urban design. The pollutants of greatest concern are those released as waste products from combustion processes via chimneys and automobiles. Chimney effluent consists largely of the nitrogen of unburned air and the main products of combustion, which are carbon dioxide, water vapour and heat. In addition there may be substantial quantities of coarse ash, fine dust and finer smoke, sulphur dioxide gas and various nitrogen oxide gases. Airborne dust eventually settles or is washed out by rain, although very fine dust may remain in the atmosphere for a few days. Plumes of chimney pollution are dispersed according to the atmospheric stability at various elevations and the height of the chimney. Dispersion upwards is almost completely prevented by even a shallow inversion layer, particularly if the chimney is relatively low.

Automobile exhausts emit lead, unburned hydrocarbons, carbon monoxide, nitrogen monoxide, water vapour, carbon dioxide and unburned nitrogen. The hydrocarbons and nitrogen oxides are precursors of photochemical pollution and react with atmospheric oxygen in the presence of adequate bright sunshine to form ozone and smaller amounts of organic gas, particularly aldehydes and the acrid gas peroxyacetylnitrate (PAN). Concentrations of ozone and PAN tend to peak in the early and mid-afternoon, a few hours after the morning maximum of traffic density (e.g. Le Roy et al., 1976; Spillane, 1978) (Fig. 6.3). Ozone reaches its highest concentrations in the warmer months when clear skies prevail. Ozone concentrations in cities like Sydney, Melbourne and London seem to be increasing and regularly exceed values recommended by the World Health Organization as levels which should not be exceeded (Ball, 1976a; Moore, 1976; Thornes, 1977b).

The visible smog which forms over major cities when traffic has been heavy, the air still and skies clear is particularly offensive. It may be due to combinations of smoke, dust, aerosols, water droplets and nitrogen dioxide. It disperses sunlight and attenuates solar radiation, reducing illumination at the ground.

Meteorological conditions clearly play an important part in determining the concentrations of pollutants. Concentrations depend on the amount of pollutant emitted into a given volume of air and on the amount of ventilation which dilutes it. The latter depends on atmospheric stability which controls convection and on the wind which controls advection. The volume into which pollutants can be diluted is determined by the depth of the mixing layer, which is fixed by the height of the lowest inversion layer. The mixing layer tends to be shallowest, and hence ground-level pollution concentrations greatest, at around breakfast time. Replenishment by daytime pollution also tends to produce another peak in late afternoon.

Valleys, which happen to be common locations for cities, are the worst places for formation and persistence of temperature inversions, because of the pooling of cold air from surrounding hills, shading from the sun which reduces convection and shelter from winds which might otherwise help to destroy inversions. On the other hand, urban heating does help to reduce the intensity of any nocturnal inversions that do occur. Persistence of haze or cloud during the daytime may reduce ground heating and hence convection. High pressure systems with light winds and subsidence inversion conditions lasting for several days may lead to progressive deterioration of conditions with prolonged accumulation of pollutants. The well-known episode in London in November 1952 was a product of just such a situation, with inversion conditions and calms for four days promoting smoke and sulphur dioxide accumulations leading to several thousand deaths before ventilation was restored.

A recent example of a bad pollution episode comes from the Greater Manchester area for the period 29 November to 1 December 1977. High pressure conditions reduced

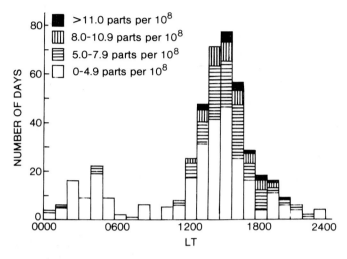

FIG. 6.3. TIME OF OCCURRENCE OF THE MAXIMUM HOURLY MEAN OZONE CONCENTRATION, ENDELL STREET, HOLBORN, LONDON, SUMMER PERIODS 1972–74
(After Stewart et al., 1976; reproduced with the permission of Macmillan Journals Ltd.).

windspeeds to about 1 m.s^{-1} or less and favoured the development of a 360 m deep surface inversion with temperatures at ground level well below freezing. The low temperatures also raised the relative humidity to saturation point, and increased demands for space heating. This led to higher than normal concentrations of smoke and sulphur dioxide, while the excess nuclei encouraged fog formation. Daily pollution levels reached peaks of nearly 700 mg.m^{-3} of smoke plus just over 700 mg.m^{-3} of sulphur dioxide at or near Manchester city centre. While such concentrations are many times less than the smogs of the early 1960s when smoke exceeded 5000 mg.m^{-3} and sulphur dioxide exceeded 3000 mg.m^{-3} they are still undesirable (Elsom, 1978).

The effects of air pollutants are many and varied. Sulphur dioxide, for example, damages metal, stonework and plants (Gillette, 1975). It can also lead to infections of the lower respiratory tract in humans, especially among the very young, the elderly or the infirm. The combination of sulphur dioxide with dust, fog, ozone or nitrogen dioxide can be particularly damaging, even crippling. The physiological consequences of pollution depend on pollution concentration and exposure time. Carbon monoxide, for example, can be expected to cause headaches when the exposure time (hours) multiplied by concentration (mg.m^{-3}) exceeds 1125. Further relationships between pollutants, particularly ozone, and health have been described in chapter 3 (see Fig. 3.9, Table. 3.5).

6.2 CLIMATE AND URBAN PLANNING

The need for urban planning to recognize effects of, and upon, climate arises at two levels of design. The first, and that which has generally received most attention, concerns the relationships between climate and architecture which are important in the design and construction of all new buildings (see chapter 3). The second is the field of urban climatology and is primarily concerned with planning to reduce urban thermal excesses and air pollution.

A basic problem for the urban planner, highlighted by Lowry (1977), is that of determining the locations, timing and intensities of the effects of urbanization on the various weather elements. In addition, the principles of urban climatology can only be applied with any great hope of success to programmes of major urban renewal or construction of new towns. In too many cases existing plans, perhaps coupled to resistance to change, impose severe constraints upon the successful application of urban climatological principles to city design (Chandler, 1976). Figure 6.4 shows one attempt to analyse the stages by which the principles of urban climatology may be introduced to the decision-making sequence, combining socio-economic and other constraints to produce the optimum climatic environment.

A city can be planned to reduce thermal stresses and the effects of air pollution. Properly spaced green areas are perhaps the most effective and aesthetically pleasing means of controlling any urban temperature excess. As well as improving the ventilation and circulatory systems of cities and thereby reducing temperature extremes, correctly planned areas of vegetation help to reduce noise levels and filter out pollutants, in addition to serving as recreation areas. The urban heat island effect could also be reduced by using building materials with lower heat conductivity and storage properties. Building materials of different compositions and/or different colours, through their effects on albedo and heat absorption might also help to create temperature differentials to encourage mixing and ventilation (Clarke and Bach, 1971). Water bodies within or close to cities help to keep maximum temperatures down and have aesthetic and recreation appeal as well (Chandler, 1976).

On the other hand, reduction of urban heat islands is not necessarily desirable in high latitude cities, which may be constructed to strengthen the heat island effect. The heat island would help to reduce space-heating demands and might substantially increase the growing season in central parks and gardens (Chandler, 1965).

The impact of air pollution can be reduced, chiefly by avoiding siting pollution sources in badly ventilated areas or in areas known to be vulnerable to frequent low inversions. Large chimneys should be built high, so that they penetrate most likely inversions, and they should be downwind from residential and business districts. The city should preferably lie in a linear fashion across the direction of prevailing winds. Planning is, however, complicated by the fact that the worst pollution conditions over a long period often accompany calm conditions or very light winds from a totally different direction. There are arguments in favour of siting major pollution sources in relation to such episodes rather than in relation to prevailing winds, which are frequently stronger, more turbulent and more efficient at reducing pollution concentrations.

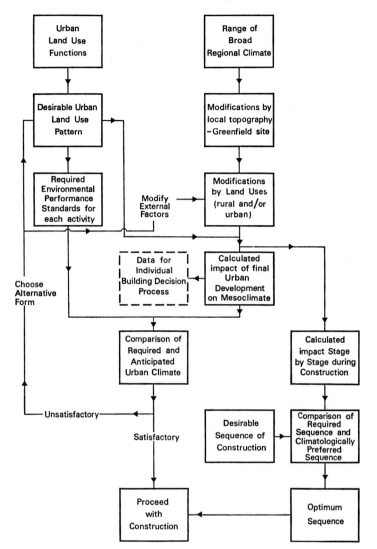

FIG. 6.4. THE SEQUENCE OF URBAN PLANNING DECISION MAKING, INCORPORATING A CLIMATIC INPUT
(After Chandler, 1976; reproduced with the permission of the World Meteorological Organization).

The relevance of climatology and meteorology to the development planning of a rapidly growing city, can be illustrated by reference to a study of Perth, Western Australia (Southern and Macnicol, 1973). The broad scale pollution climatology of the Perth area was assessed from a detailed study of low level wind regimes and associated stability patterns in the form of mixing depth statistics. A major feature of Perth's climate, especially in the winter months is the predominance of early morning east to north-east land breezes, which points to preferred industrial location to the south-west or north-west of the Perth city area, with potentially harmful pollution concentrations being directed out to sea. The mixing layer on such occasions may extend to 1200 m, but is usually below

600 m. During daylight hours the situation is reversed. Perth is the sunniest and windiest Australian State capital, so turbulent motion aids effective pollution dispersal. Mixing depths during the day are substantially deeper than at night, most commonly being between 1200 and 1800 m, sometimes greater. Coupled with the usually strong sea breezes, particularly in summer, high turbulent motion produces an erratic variation in the downwind ground level pollution concentrations. Since most daytime winds are from the south-west or south-east, preferred industrial locations would be to the north. A combination of mixing depth and prevailing wind velocities thus indicates that both under early morning conditions of high stability and afternoon erratic dispersion, an industrial area located to the north-west of Perth would produce a minimum number of pollution occurrences over the city and environs.

The appropriate location of an airport providing convenient access to a city, while reducing the incidence of aircraft noise and pollution is of obvious importance in planning a city and was also considered by Southern and Macnicol (1973). In this case major attention was focused on questions of safety and noise, although studies elsewhere, such as that of Honolulu International Airport, have looked more closely at air quality (Daniels and Bach, 1976). The siting of Perth's major civil and military aerodromes so near to the Darling Range just east of the city can be criticized because of the effects of relief on airstreams flowing over the range. The forced ascent of moist air, such as Perth's winter westerlies, for example, enhances low base cloudiness and shower activity on the windward slopes of the range. In such situations, particularly during the passage of cold fronts or low pressure troughs, airport visibility may deteriorate below the acceptable minima for aircraft operations. The minimum operating conditions for Perth airport are a cloudbase of 181 m (242 m at night) and a visibility of 3700 m. If the forecast cloudbase exceeds these heights by less than 152 m, approaching aircraft are required to carry additional fuel in case diversion to an alternative airport is necessary. The siting of an airport in a locally weather-sensitive area thus reduces the economic operation of air routes which include that location.

The Darling Range east of Perth also has the effect of creating severe low-level air turbulence under certain well-defined meteorological conditions. This hazard is most noticeable in summer when a stable easterly or north-easterly airstream blowing across the range is reinforced by a land breeze and by katabatic winds flowing down the slopes as a result of differential overnight cooling. Experienced pilots have described Perth Airport as the most unsuitable in Australia for operations in such gusty conditions. Again, a site northwest of Perth could have alleviated such problems and because of the high incidence of prevailing south-west and north-east winds would also have facilitated aircraft approaches along routes less likely to create a noise problem for the residential areas of the city.

The whole question of airport siting is a very difficult one for planners to solve. On a much smaller scale the airport at Armidale in New South Wales, the highest commercial airport in Australia at an altitude of just over 1000 m, suffers similar weather-related operating problems. Low cloud bases frequently close the airport to all traffic, necessitating diversions and delays for passengers and increased operating costs for airlines. In addition, the airport is situated at the top of a hill above the small city of Armidale and regularly suffers from fog at about 0900 h on winter mornings, which just happens to be the departure time for the daily flight to Sydney. The problem, unforeseen by planners in this instance is that fog accumulates in the valley over the city during cold, clear and calm winter nights, but tends to drift upslope towards the airport and higher residential areas during the early morning. Such conditions probably apply in many other parts of the world and the examples discussed simply serve to illustrate the sorts of things that planners should consider.

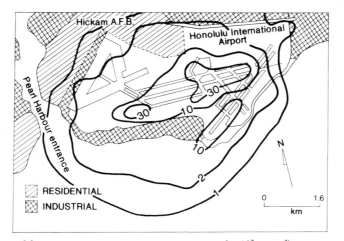

FIG. 6.5. CALCULATED 3 h MAXIMUM HYDROCARBON CONCENTRATIONS ($\times 10^2 \mu g.m^{-3}$) AT HONOLULU INTERNATIONAL AIRPORT IN 1971/72
(After Daniels and Bach, 1976; reproduced with the permission of the Air Pollution Control Association).

Reference has already been made to the study by Daniels and Bach (1976) of the environmental impact of Honolulu International Airport on the surrounding air quality. It is worth looking at this in greater detail as a further example of the questions to which planners should be addressing themselves. The accurate assessment of temporal and spatial pollutant emission patterns at airports involves many variables: air traffic density by type, and engine make and model of aircraft; the use of fuel by different aircraft; the pollutant emission rates by engine model and operational mode; the allocation of emission rates to the respective runways, turn-off points, taxi-ways and parking areas; and the time each aircraft spends in the different operational modes. The resulting emission patterns (Table 6.2, Fig. 6.5) then have to be related to meteorological information. Daniels and Bach calculated maximum surface concentrations of emissions and compared them with local ambient air quality standards (Fig. 6.6). They demonstrated that even in 1971/72 the local standards proposed to come into effect in 1975 were already being substantially exceeded. It is clear that major airports may have a considerable adverse impact on the surrounding air quality, and this element needs to be borne in mind at all stages of the planning process (see also, George et al., 1972).

As with pollution from other urban sources, air pollution from automobiles can be reduced in a number of ways. The most obvious and relatively common approach is to control emissions either through the fuel or through exhaust abstraction systems, although the efficiency of such systems is debatable and public acceptance of their use is often poor.

Table 6.2

ANNUAL AMOUNTS OF AIR POLLUTANTS EMITTED AT HONOLULU INTERNATIONAL AIRPORT FROM AIRCRAFT AND GROUND OPERATIONS, 1971/72
(after Daniels and Bach, 1976)

Source	Number	Air pollutants (t.a^{-1})				
		Particulate matter	Carbon monoxide	Nitrogen oxides	Comb. organic gases	Sulphur dioxide
Commercial jet aircraft	115 443	1409	2 922	1 390	5 646	425
Commercial and private piston aircraft	122 008	25	6 076	287	1 135	negl.
Military jet aircraft	44 986	564	1 106	386	689	157
Military piston aircraft	8 978	58	13 978	661	2 611	negl.
Total aircraft operations	291 415	2056	24 082	2 725	10 081	582
Aircraft fueling systems	62				25	
Aircraft engine run-ups	3 384	24	113	7	136	6
Vehicles entering and leaving airport	12 175 670	5	1 184	79	215	4
Operation of service vehicles	677	52	11 073	522	2 361	31
Total ground operations		81	12 370	608	2 737	41
Grand total		2137	36 452	3 333	12 818	623

The implementation of air pollution control measures in respect of new passenger cars in Australia has been estimated to have cost about A\$ 100 per vehicle at 1976 cost levels.

Planners can make a positive contribution through the design and layout of city streets and buildings to promote greater turbulence and better diffusion of pollutants. In general good traffic flow helps to promote better air quality in urban areas, but traffic engineering

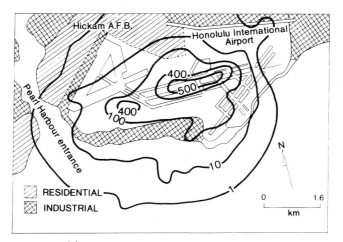

FIG. 6.6 CALCULATED NUMBER OF 3 h PERIODS PER YEAR AT HONOLULU INTERNATIONAL AIRPORT WITH HYDROCARBON CONCENTRATIONS ABOVE THE HAWAII AIR QUALITY STANDARD FOR ONE 3 h PERIOD (11 μg.m^{-3})
(After Daniels and Bach, 1976; reproduced with the permission of the Air Pollution Control Association).

and air pollution control aims are frequently at variance (Patterson, 1975). One aspect of traffic engineering which might help to improve the situation is the construction of bicycle tracks to encourage bicycle riding. Unfortunately, the situation in many cities at the moment means that such aims are incompatible, since while increased bicycle numbers at the expense of automobiles should help to reduce emissions, present and immediate future emission levels are high enough to act as positive deterrents to bicycle riders (Kleiner and Spengler, 1976; Wright, 1975).

The implementation of fuel conservation policies, use of alternative types of fuel, and new pollutant emission regulations are steps towards simplification of the basic urban planning problem, through minimization of total pollution emissions. On the other hand it can be argued that it is better and cheaper to achieve clean air from the outset through sensible planning than to have to take expensive remedial steps later on (Sherwin, 1974). Evidence from the UK tends to support the role of the Clean Air Acts of 1956 and 1968 in reducing smoke concentrations, although in some places reductions seem to have occurred without the application of the provisions of the Act (Weatherley, 1974). By 1977 about 5000 smoke control orders had been confirmed covering more than 7 million premises.

The United States Clean Air Act of 1970 placed responsibility upon the States for assuring air quality within their borders. Under this Act, which amended earlier legislation of 1963, 1965 and 1967, the Environmental Protection Agency sets national ambient air quality standards for oxides of sulphur, oxides of nitrogen, carbon monoxide, photochemical oxidants, hydrocarbons, particulates, toxic metals, other hazardous substances, odours

Table 6.3
EXAMPLES OF AIR POLLUTANT EMISSION AND AMBIENT AIR QUALITY STANDARDS
(after Gilpin 1978; Perkins 1974)

Pollutant	Australian emission standard	Averaging time	USA ambient air quality standards		
			California	Federal Primary[a]	Federal Secondary[b]
Smoke	Ringelman 1[c]	1 observation	Sufficient to reduce prevailing visibility to 10 miles with RH < 70%		
Solid particles	0·1 to 0·5 g.m^{-3}[d]	24 h	100 μg.m^{-3}	260 μg.m^{-3}	150 μg.m^{-3}
Carbon monoxide	1·0 g.m^{-3}	12 h			
		8h	10 p.p.m. (11 mg.m^{-3})	9 p.p.m. (10 mg.m^{-3})	9 p.p.m. (10 mg.m^{-3})
		1 h	40 p.p.m. (46 mg.m^{-3})	35 p.p.m. (40 mg.m^{-3})	35 p.p.m. (40 mg.m^{-3})
Nitrogen dioxide	0·5 to 2·0 g.m^{-3}[e]	Annual		0·05 p.p.m. (100 μg.m^{-3})	0·05 p.p.m. (100 μg.m^{-3})
		1 h	0·25 p.p.m. (470 μg.m^{-3})		
Sulphur dioxide		Annual		0·03 p.p.m. (80 μg.m^{-3})	0·02 p.p.m. (60 μg.m^{-3})
		24 h	0·04 p.p.m. (105 μg.m^{-3})	0·14 p.p.m. (365 μg.m^{-3})	0·10 p.p.m. (260 μg.m^{-3})
		3 h			0·5 p.p.m. (1300 μg.m^{-3})
		1 h	0·5 p.p.m. (1300 μg.m^{-3})		
Hydrogen sulphide	5 mg.m^{-3}	1 h	0·03 p.p.m. (42 μg.m^{-3})		
Lead	10 mg.m^{-3}	30-day	1·5 μg.m^{-3}		
Chlorine	1·0 g.m^{-3}				

[a] National Primary Standards: levels of air quality necessary, with an adequate margin of safety, to protect the public health

[b] National Secondary Standards: levels of air quality necessary to protect the public welfare from any known or anticipated adverse effects of a pollutant

[c] Visible pollution can be assessed approximately by means of matching against four patches of greyness on a card called a Ringelman Chart

[d] 0·1 g.m^{-3} (furnaces for heating metals); 0·25 g.m^{-3} (solid fuel boilers, incinerators > 300 kg/h, any other trade, industry or process); 0·5 g.m^{-3} (incinerators < 300 kg/h)

[e] 0·5 g.m^{-3} (any trade, industry or process except nitric or sulphuric acid plants or gas-fired power stations); 1·0 g.m^{-3} (sulphuric acid plants); 2·0 g.m^{-3} (nitric acid plants).

and noise. The aim has been to reduce motor vehicle pollution by 90 per cent, while requiring the best available control technology for all new pollution sources (Table 6.3). In the electricity generating industry, pollution control equipment, such as chimneys of advanced design, and dust-arresting and ash disposal equipment may add 6 per cent to the cost of a new coal-fired power station (Gilpin, 1978). In Los Angeles county exceptionally high expenditures have been incurred by industry in meeting the very severe restrictions on industrial emissions. The cost of control equipment averages about 25 per cent of the cost of basic production equipment. Such costs have to be balanced against estimated average annual costs of air pollution. The United States Environmental Protection Agency in 1972 estimated that the average annual cost of air pollution was about US\$ 65 per person (see Table 6.4).

Table 6.4
SOME ESTIMATES OF THE ANNUAL COSTS OF AIR POLLUTION IN THE UNITED STATES
(after Gilpin, 1978)

Source	Base Year	Estimate (US$ billions)
Gerbardt	1968	6·0 to 15·2
Barrett and Waddell	1968	16·1
Babcock and Nagda	1968	20·2
Ridker	1970	7·3 to 8·9
Justice, Williams, and Clement	1970	2·0 to 8·7
Waddell	1970	6·1 to 18·5
National Academy of Sciences	1973	15·0 to 30·0
Heinz and Herschaft	1973	9·5 to 35·4

The World Health Organisation has been involved in air pollution control since the mid-1950s (Cleary, 1974). On the basis of epidemiological and other evidence of effects on man and his environment of a range of atmospheric pollutants the WHO proposed air quality levels involving varying degrees of health risk for each of the pollutants. Air quality levels were established for sulphur oxides and suspended particulates taken together, for carbon monoxide and photochemical oxidants. In addition, long term goals were put forward for national air pollution standards (see Table 6.3). The difficulties in meeting such standards are clearly illustrated by the case of Sydney, where in 1975 the WHO carbon monoxide goal over 8 hours was exceeded on 174 days and the ozone goal on 233 days. Similarly, ozone measurement in London during the hot sunny summers of 1975 and 1976 showed a peak hourly concentration of 15 and 21 p.p.h.m. respectively, far exceeding the WHO recommended maximum concentration of 8 p.p.h.m.

A very important consideration which is just starting to gain recognition is that the air pollution problem is no longer confined, if in reality it ever was, to the local scale. Recognition of the importance of long distance transport of air pollution, particularly in relation to ozone, means that the problem is becoming one that needs to be tackled by legislators at the continental scale (e.g. Cox et al., 1975; Fankhauser, 1976; Thornes, 1977b).

A useful approach to urban planning is through computer modelling of the dispersion of pollutants from existing and potential sources. It is possible to enter data on emission rates, chimney heights, atmospheric stability and the directions, strengths and frequencies of local winds, and then to calculate average ground level pollution concentrations anywhere in the region. Thus vulnerable areas can be identified and the effects of alternative locations of proposed pollution sources can be tested. Schwerdtfeger and Lyons (1976) used this sort of approach for a study of the potential distribution of possible pollutants over a 500 km^2 area of urban and suburban Adelaide.

An important starting point for such studies is the compilation of an air pollution emission inventory to establish the sources of air pollution, the type and quantity of pollutants emitted, and the emission characteristics of each source. Inventories have been established and in many cases used in models for many of the larger urban areas in northern Europe, the USA, Canada, Japan and Australia, but relatively few attempts have been made in the UK. Ball (1976b) discusses such an inventory for London, and points to the need for many more similar studies in the large British conurbations.

SUGGESTIONS FOR FURTHER READING

ALLEN, G. H. and IVERACH, D., 1976; 'Some comments on a high ozone day.' *Clean Air*, 6, pp. 23–34.

ANON, 1978, 'Tackling Sydney's smog.' *Ecos*, 16, pp. 3–9.

ATWATER, M. A., 1975, 'Thermal changes induced by urbanisation and pollutants.' *J. Appl. Met.*, 14, pp. 1061–71.

BALL, D. J. and BERNARD, R. E., 1978, 'Evidence of photochemical haze in the atmosphere of Greater London.' *Nature*, 271, pp. 733–4.

CECH, I., WEISBERG, R., HACKER, C. and LANE, R., 1976, 'Relative contribution of land uses to the urban heat problem in the coastal subtropics.' *Internl. J. Biomet.*, 20, pp. 9–18.

CHANDLER, T. J., 1964, 'City growth and urban climate.' *Weather*, 19, pp. 170–1.

—1976, 'The Royal Commission on environmental pollution and the control of air pollution in Great Britain.' *Area*, 8, pp. 87–92.

CHANGNON, S. A., 1969, 'Recent studies of urban effects on precipitation in the United States.' *Bull. Amer. Met. Soc.*, 50, pp. 411–21.

CRADDOCK, J. M., 1965, 'Domestic fuel consumption and winter temperatures in London.' *Weather*, 20, pp. 257–8.

DETWILER, A. 1974, 'Urban influence on cumulus formation.' *Bull. Amer. Met. Soc.*, 55, pp. 1240–1.

FERRARI, L. M. and JOHNSON, D. C., 1976, 'Photochemical smog—the Sydney scene.' *Clean Air*, 10, pp. 2–9.

FINDLAY, B. F., 1976, 'Meteorology as related to urban and regional land-use planning.' *WMO Bull.*, 25, pp. 175–7.

FREESTONE, D. H., 1974, 'Wind environment of buildings.' *Town Planning Quarterly*, 35 pp. 34–7.

GARNETT, A. and BACH, W., 1965, 'An estimation of the ratio of artificial heat generation to natural radiation heat in Sheffield.' *Mon. Weath. Rev.*, 93, pp. 383–5.

HANNA, S. R., 1977, 'Predicted climatology of cooling tower plumes from energy centres.' *J. Appl. Met.*, 16, pp. 880–9.

HANNA, S. R. and GIFFORD, F. A., 1975, 'Meteorological effects of energy dissipation at large power parks.' *Bull. Amer. Met. Soc.*, 56, pp. 1069–76.

HANNELL, F. G., 1976, 'Some features of the heat island in an equatorial city.' *Geografiska Annaler*, A, 58, pp. 95–110.

INHABER, H., 1975, 'A set of suggested air quality indices for Canada.' *Atmos. Env.*, 9, pp. 353–64.

JONES, P. M. and WILSON, C. B., 1968, 'Wind flow in an urban area: a comparison of full scale and model flows.' *Building Sci.*, 3, pp. 31–40.

LANDSBERG, H. E., 1974, 'The urban area as target for meteorological research.' *Bonner Meteorologische Abhandlungen*, 17, pp. 475–80.

LEE, D. O., 1977, 'Urban influence on wind directions over London.' *Weather*, 32 pp. 162–70.

LUCAS, D., 1974, 'Pollution control by tall chimneys.' *New Scientist*, 63, pp. 790–1.

MAINWARING, S. J. and HARSHA, S., 1976, 'Size distribution of aerosols in Melbourne city air.' *Atmos. Env.*, 10, pp. 57–60.

MENZIES, J. B., 1971, 'Wind damage to buildings.' *Building*, 221, pp. 67–76.

MOORE, F. K., 1976, 'Regional climatic effects of power plant heat rejection.' *Atmos. Env.*, 10, pp. 805–12.

PASQUILL, F., 1972, 'Factors determining pollution from local sources in industrial and urban areas.' *Met. Mag.*, 101, pp. 1–8.

PATTERSON, R. M., 1975, 'Traffic flow and air quality.' *Traffic Engineering*, 45, pp. 14–17.

ROSS, F., 1974, 'Fuel conservation beats air pollution.' *Surveyor*, 143, pp. 46–9.

SPURR, G., 1974, 'Meteorology and cooling tower operation.' *Atmos. Env.*, 8, pp. 321–4.

TAUBER, S., 1976, 'A systems approach to air pollution control.' *Atmos. Env.*, 10, pp. 633–6.

TERJUNG, W. H., 1970, 'The energy-balance climatology of a city-man system.' *Annals Assoc. Amer. Geogrs.*, 60, pp. 466–92.

TERJUNG, W. H. and LOUIE, S., 1973, 'Solar radiation and urban heat islands.' *Annals Assoc. Amer. Geogrs.*, 63, pp. 181–207.

TORRANCE, K. E. and SHUM, J. S. W., 1976, 'Time-varying energy consumption as a factor in urban climate.' *Atmos. Env.*, 10, pp. 329–38.

TRAIN, R. E., 1974, 'Pursuing the goals of the Clean Air Act.' *J. Air Polln. Control. Assoc.*, 24, pp. 740–51.

TULLER, S. E., 1975, 'The energy budget of man: variations with aspect in a downtown urban environment.' *Internl. J. Biomet.*, 19, pp. 2–13.

WALL, G., 1974, 'Complaints concerning air pollution in Sheffield.' *Area*, 6, pp. 3–7.

7
Weather Forecasting

The emphasis so far has been upon the ways in which the atmosphere impinges upon man's activities, with some attention to the adjustments he makes in order to cope with the conditions presented. In this chapter and in chapter 8 attention is turned to the positive responses that man employs in efforts to use available atmospheric resources to the fullest extent. Several references have been made in the preceding chapters to weather forecasting, to the diverse demands for forecasts and to some of the problems involved. These are now examined in greater detail.

7.1 THE FORECASTING PROBLEM

The preparation of weather forecasts for public and private interests is one of the most important activities in the field of meteorology (American Meteorological Society, 1976a). Many individuals use weather forecasts to guide their daily living patterns, and agriculture, industry, commerce, and government use them to guide their operations. Such widespread needs for accurate advice on expected weather conditions and the often critical dependence of public safety and welfare upon the quality of such information highlight one of the basic problems of the forecaster.

The wide range of consumers of weather forecasts implies a similarly broad range of requirements and expectations of the forecasts. Requirements may vary in terms of the actual weather parameters desired in the forecast, and in terms of time and space scales to be specified. It is impossible, for example, to provide detailed and accurate weather forecasts for all users at all locations over a large area. The scale of forecasting usually means that reliable regional forecasts are possible but that such forecasts will not necessarily be accurate for all points within the region; local variations are inevitable. For example, a forecast of scattered showers will probably be correct on the regional scale, but it is impossible to specify where the showers will be, when they will occur, how long they will last or how much rain they will bring. Similarly, it may be possible to forecast with reasonable certainty that a region will experience general rain, but there will usually be some parts that do not get any rain at all, and it is again impossible to say how much, where or when rain will fall. In all such situations the forecasts will be inaccurate for many people.

The weather forecaster is still faced with several constraints upon his ability to please everybody. The laws of physics governing the weather include a random element, such as the effects of man's combustion of fossil fuels. The variability of such inputs to the atmospheric system considerably complicates the forecasting process. Indeed, as indicated in chapter 1, our understanding of the complexity of atmospheric circulation is still far from perfect, and it is this which lies at the root of the forecasting problem. Not only do we have an incomplete understanding of the working of the atmosphere, but the forecaster is also hampered by a far from complete knowledge of present atmospheric conditions to use as a starting point in estimating the future. Many large areas of the world, especially over the oceans, major desert and mountain regions, in low latitudes, and in the southern hemisphere suffer from lack of basic data because of very coarse collecting networks.

Ramage (1976) has suggested that unless observing systems are designed to pick up relatively unpredictable turbulence 'bursts' forecasting will not improve. He based this conclusion on the hypothesis that unpredictable turbulence bursts, which accomplish most mixing and transfer within fluids (in this case the atmosphere), are the most important agents of atmospheric change and may set a limit to atmospheric predictability. Amongst other measures indicated to improve the situation is the improvement of the observing network, incorporating finer temporal and spatial resolutions.

Unfortunately, an increase in the number of observation points through the three-dimensional atmosphere, sufficient to meet the requirements for very localized forecasts, could in itself create new data handling problems. It is conceivable that there would be so much data to incorporate in the forecast procedure that it might be impossible to produce a forecast for say 12 hours ahead in less than 12 hours. So, there would seem to be a balance point between the desired levels of detail and accuracy possible in forecasts and the physical data handling problems, to permit production of a forecast with sufficient lead time to the forecast events.

The problems of knowledge of the working of the atmosphere, of temporal and spatial scale variability, of unpredictability and of data availability, while considerable for short-term forecasting up to 24 hours ahead become almost insurmountable as the length of forecast increases. In many respects, particularly in the cases of agricultural planning, construction operations, or retailing, long-term forecasts for a month or even six months ahead are more useful than daily forecasts. The problems involved in preparing an accurate daily forecast become compounded, however, in the long term, and the degrees of precision and reliability inevitably decrease.

The magnitude of the problem involved in accurate long period forecasting is well illustrated in the following example for rainfall. The average time for water to cycle between the atmosphere and the earth's surface is about nine days (chapter 1). Thus accurate prediction of rainfall amounts for a month ahead may be regarded as akin to tracing the paths of individual air particles and calculating the details of their passage, three times, through the evaporation-precipitation cycle. Lorenz (1969, 1973) has shown that on a day-to-day time scale the limit of predictability is about twelve days, and Ratcliffe (1978) reported this to be substantiated by work in Britain. He also expressed scepticism about the prospects for climatic forecasting on a time scale of decades, centuries or longer, the problems depending to some extent on man himself.

The difficulties facing the weather forecaster may be summarized as falling into three general categories: physical, mathematical, and observational. The basic problem for weather forecasting stems from the fact that the meteorologist is faced with determining the net result of a wide variety of interacting physical processes over scales from millimetres to thousands of kilometres and from seconds to centuries.

7.2 TYPES AND METHODS OF FORECASTING

An elementary form of weather forecasting can be seen in folklore, mythology and superstition, which are full of references to weather events. Many folklore sayings seem to be based on fairly sound principles, although these are probably unrealized, but just as many beliefs are quite wrong. The transition from folklore to science is connected with the development of instruments (thermometer by Galileo, 1593; barometer by Torricelli, 1643); with the accumulation and mapping of observations (e.g. a wind map by Halley, 1686, and a weather map by Buys Ballot, 1852); and with the development of new concepts such as the Bjerknes cyclone model of 1920. Improved communications have been a very important element in the development of scientific weather forecasting, facilitating rapid accumulation of data from large areas. Further impetus has been given to weather forecasting through the recent developments in the fields of radar, satellite, and computer technology, through the growth of international collaboration in research and exchange of information (see chapter 8), and through the development of mathematical theory and models.

At the personal level weather forecasts can be attempted on the basis of observation of the immediate surroundings. This may incorporate the current appearance of the sky and recent changes, local experience and perhaps information from the news media. While many people may prefer to rely on such methods, the majority place at least some reliance on the professional forecasting services. Three basic methods of forecasting can be identified: use of the statistics of past records, on the assumption that the future will be a repetition; synoptic forecasting based on the derivation of trends from regular measurements of the weather by networks of observers and incorporating data from other sources such as satellites; and numerical forecasting which uses the basic principles of atmospheric physics and large, high-speed computers to calculate how conditions will change from an initial state based on observation.

The simplest method of forecasting from past statistics is to take the average value of a weather element like temperature, for a particular date, time and place and to use it as a guide to future conditions. An alternative approach is to identify regularities or periodicities in the past and to assume their continuation in the future. Unfortunately many periodicities have been reported but few have proved reliable for predictive purposes. One atmospheric periodicity that does seem to have some potential for seasonal forecasting, perhaps up to a year ahead is the quasi-biennial oscillation (QBO) in equatorial/stratospheric winds (Holton and Lindzen, 1972; Ebdon, 1975). Ratcliffe (1978) reported some success with short-term climatic forecasting in both the northern and southern hemispheres using knowledge of the phases of the QBO and saw prospects for increasingly reliable forecasts a year or so ahead.

Another periodicity that may prove useful for seasonal forecasting is the Southern Oscillation (chapter 2), which is a fluctuation of the atmospheric circulation with an irregular period. The Southern Oscillation is related to variations in the strength of the

Walker Circulation, which is a circulation in the plane of the equator involving ascent over the Indonesian sector and subsidence over the eastern Pacific. When the Walker Circulation is strong, sea-surface temperatures are low in the equatorial eastern Pacific. Rainfall and cloudiness are also low in the same areas but high over Indonesia and Australia, and other associated anomalies of weather and circulation occur in many parts of the world. The strength of the Walker Circulation fluctuates irregularly but with a characteristic period of two to five years. If the mechanisms of the Southern Oscillation could be understood and various lag relationships between it and air temperature, sea-surface temperatures and rainfall could be identified prospects seem good for seasonal forecasting up to two years ahead (Wright, 1975, 1978).

Synoptic weather forecasting is the type with which the general public is most familiar through daily newspapers, radio and television. It starts with the production of a map of the current weather situation, followed by prognosis of a similar map for 24 hours or some other time later. The prognostic chart is interpreted in terms of the weather elements, particularly temperatures, rainfall and winds. The map of current weather is compiled from the numerous data gathered from surface observations, from radiosonde determinations of winds, temperatures and humidities at various heights, from meteorological satellites, radar, ships and aircraft (see Anderson, 1978, for a very useful discussion of current uses of satellite data in forecasting). Weather data are shared internationally through the World Weather Watch (WWW) system, inaugurated by the World Meteorological Organization in 1965. The main data handling stations are at Washington, DC, Melbourne, and Moscow, with major regional links at places such as Prague, Brasilia and Nairobi (Fig. 7.1) (see chapter 8 for further details of the WWW).

Personal skill and experience are still important elements in the synoptic forecasting process, so that accuracy is restricted by human limitations. In the interests of greater economy and speed of prognosis, together with the need to handle growing amounts of input information from an increasing variety of data sources, increasing use is now being made of numerical forecasting. This involves the construction of models designed to reproduce the basic properties of the atmosphere. Several important considerations are involved in the modelling (Miyakoda, 1974; Shuman, 1978). The whole system of equations has to be simplified as much as possible, depending upon the accuracy required in the forecasts and on the range of applicability. Meteorologically unimportant components such as acoustic waves are filtered out. Secondly, the system of equations must be designed to guarantee the conservation of air mass, water vapour, momentum, and total energy for the entire volume for all time. Thirdly, the spatial resolution must be selected carefully. In principle, the finer the grid resolution the better, but computer limitations impose restrictions. In addition, the mathematics themselves are complicated and have to deal with four-dimensional fields of data.

The physical assumptions adopted and the degree of sophistication of each physical process incorporated differ from model to model. Numerical forecasting is still accompanied by numerous problems arising from the need for adequate data to define the initial condition of the atmosphere, from the problem of determining and allowing for changes in conditions at the boundary of the model, and from the random nature of the eddies of various sizes which are inherent in atmospheric motions. Any errors in prediction increase each time the calculations are repeated after a time-step. Several billion arithmetic and

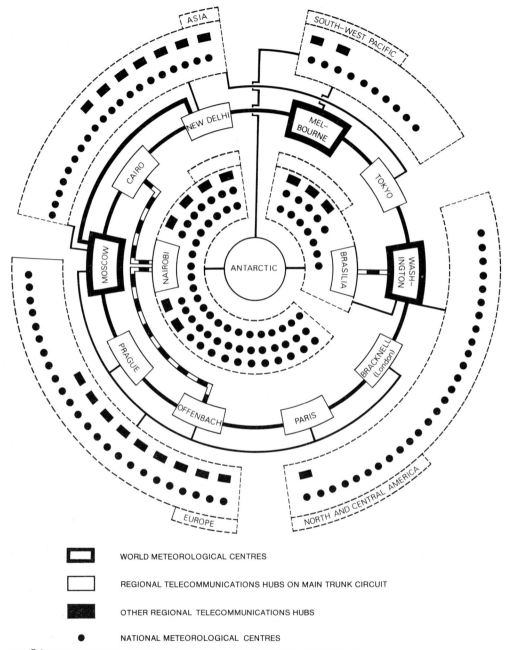

FIG. 7.1. SCHEMATIC DIAGRAM OF THE WMO GLOBAL TELECOMMUNICATIONS SYSTEM
(After Wallen, 1974; reproduced with the permission of the World Meteorological Organization).

logical operations have to be performed to complete one numerical prediction with current models, and operational deadlines must be met. This requires the fastest computers on the market (Fig. 7.2).

In practice, the weather forecaster uses a mixture of conventional synoptic forecasting procedures with guidance from numerical products. Kirk (1974) saw the forecaster's essential role as the exercise of discrimination or judgement within the context of the

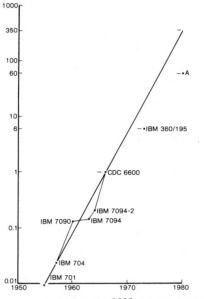

FIG. 7.2. OPERATIONAL COMPUTER SPEED, RELATIVE TO THE CDC-6600, PLOTTED AGAINST YEAR OF ACQUISITION FOR NMC USE
(After Shuman, 1978; reproduced from the *Bulletin of the American Meteorological Society*, p. 9, with the permission of the American Meteorological Society).

available information and the needs of his customers. He has available to him a variety of numerical products based on different models and different scales, from many centres around the world; he has observations from observing networks, including satellites and radar; he has special knowledge about the data deficiencies of any particular model and about model deficiencies in terms of performance limitations or possible errors; and he has a background of theoretical and empirical knowledge. Judgement may be exercised depending upon the needs of a particular situation. In essence, objective methods should be used to the fullest extent compatible with forecast requirements, yet permitting the application of theoretical knowledge and information that the model cannot handle. The result, as in operational procedures at the Central Forecasting Office in Bracknell, is a complex mix of subjective and objective inputs to the forecast.

Many of the severe weather phenomena most important to human activity occur on space scales smaller than those that are resolved by current operational numerical models, so considerable efforts are being devoted to the development of numerical models capable of predicting severe storms at the mesoscale (Anthes, 1976). Mesoscale severe weather phenomena are not restricted to tornadoes and thunderstorms, and may include heavy snow and ice storms, hurricane force downslope winds, floods, and severe air pollution episodes. Many such mesoscale systems are associated with terrain variations, which introduce further complexity to the models required.

The uncertainties in forecasts up to 24 h ahead become compounded as the length of the forecast period grows, yet the attractions of accurate long range weather forecasts are obvious for many areas of economic activity. Figures from the USA show that about 2000 customers subscribe on a regular basis to the monthly forecasts issued by the long range weather forecasting office in that country. These customers are spread across a wide range of activities, most of them, surprisingly, outside the 'agribusiness' sector. This may be related to the fact that only 20 to 30 per cent of the variability in major crop yields in the USA can be attributed to meteorological factors. It can be argued that even forecasts of limited predictive skill, which probably means most long range weather forecasts, could be of some value to those who knew how to use them. Investigations in England and the USA have shown the possibility of forecasting for a week or two ahead with an above-chance success on the basis of current deviations from normal. Work by Ratcliffe (1974), Bell (1976) and Lee and Ratcliffe (1976), for example, has demonstrated the usefulness of 500 mbar anomalies, northern hemisphere circulation anomalies, and surface pressure anomalies respectively, for forecasts ranging from 15 days to a season. Much long range forecasting is based on numerical modelling, but the use of analogues has yielded useful results.

One of the most promising areas for the development of long range forecasting seems to be in the establishment of relationships between the atmosphere and the oceans, particularly in view of the relatively conservative properties of the latter. Sea surface temperature is an important influence on the atmospheric circulation, but the interrelationships are complicated by the fact that the atmosphere itself reacts on the ocean (Wright, 1977).

Other approaches to the problem, particularly on the time-scale extending to a year or even many years ahead include statistical analyses, such as that by Dyer and Tyson (1977) for estimating above and below normal rainfall periods over South Africa up to the year 2000, and the establishment of the predictive significance of solar-climatic cycles (Willett, 1974).

Laurmann (1975) has illustrated the dimensions of the seasonal climate forecasting problem by listing the interacting submodels that make up the total climate system. In essence these are the same features that apply to forecasting on any time scale. These submodels include the troposphere, the stratosphere, the planetary boundary layer, the evaporation-precipitation cycle, description of the atmospheric radiative heat flux distribution, air–sea interaction, a shallow mixed layer ocean model, the general circulation of the oceans, surface heat transfer and radiative processes, the cryosphere and its coupling with the oceans and the atmosphere, ice dynamics, and solar flux input to the globe. Simplified representations are required for the processes in these sub-models, but there are still considerable inadequacies in the ability of meteorologists to achieve satisfactory approximations. All weather and climate forecasting models make gross simplifying assumptions which may in fact not be justified. Many basic research questions remain to be answered before a satisfactory seasonal forecasting model is possible.

Seasonal weather forecasting is not the exclusive domain of professional meteorologists employed by the various national weather services. In Australia, for example, Lennox Walker has clients ranging from many of the major business organizations, through farmers and graziers, fete organizers, politicians and film producers to brides. His method is based on sunspot activity and reference to early weather records from around the world. Most of his forecasts cover weather patterns for about 18 months ahead but he does make predictions up to 15 years in advance for clients planning special long range projects. There is no doubt that Lennox Walker has made a very successful career out of his forecasting activities and he seems to have a regular clientele, but it is probably true to suggest that his methods are still viewed with some scepticism by the professional meteorologists of the Bureau of Meteorology.

7.3 ACCURACY AND RELIABILITY OF FORECASTS

The literature on weather forecasting contains many references to and investigations of the skill of forecasters and the accuracy of their forecasts. This seems reasonable, since we clearly require some yardstick against which to judge the forecasts we see or hear. In this section, therefore, a brief review is made of some of the evaluations of forecasting skill and accuracy, as indicators of the levels being attained.

Skill in weather forecasting varies with the geographical area, the meteorological situation, the season, and with the length of the period over which the forecast is projected. Forecast accuracy can be determined objectively by comparison with simple, derived estimates using the principles of persistence or of climatology. The use of persistence entails predicting that the weather will remain unchanged, while application of climatological principles relies on the prediction of average weather based on past records. Skill cannot be said to exist in a forecast unless the accuracy exceeds levels achieved by basic methods such as these (American Meteorological Society, 1976a).

Figure 7.3 shows the history of skill at the National Meteorological Center (NMC) in Washington, DC, with 36 h circulation predictions at the 500 mbar level. During the 20 years of numerical weather prediction from 1955, skill can be seen to have increased from about 30 per cent to about 60 per cent according to the measuring scale employed. The measure of skill is based on the so-called S_1 score, which is a measure of normalized error in horizontal pressure gradients. A chart with an S_1 score of 20 is virtually perfect, and one with 70 is worthless. Skill (per cent) is $2 \times (70 - S_1)$, which yields 0 for a worthless chart and 100 for a virtually perfect one (Shuman, 1978). Predicted circulation patterns at other levels in the atmosphere also increased in skill.

The results of 24 h operational predictions for three levels of broad-scale wind patterns over the Australian region using conventional synoptic methods and two models are shown in Figure 7.4. The relative accuracy of predictions is indicated by a skill score based on the vector difference between predicted and observed winds. The lower the skill score the more accurate the prediction. The skill scores of persistence forecasts are included for comparison.

Temperature is one of the easiest weather elements to forecast, but precipitation, particularly quantitative precipitation, is one of the most difficult. Useful forecasts of maximum and minimum temperatures at individual American cities have been provided to

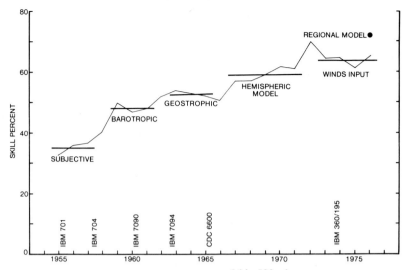

FIG. 7.3. RECORD OF SKILL, AVERAGED ANNUALLY, OF THE NMC 36 h, 500 mbar PREDICTIONS OVER NORTH AMERICA (After Shuman, 1978; reproduced from the *Bulletin of the American Meteorological Society*, p. 12, with the permission of the American Meteorological Society).

forecasters from NMC's computers for several years (Fawcett, 1977). A steady increase in skill from 1968 to 1975 is shown in Figure 7.5. Forecasts of whether or not rain will occur have been kept since 1954 for Boston, Chicago and Washington. These are clearly the easiest form of precipitation forecast to make and have shown steady improvement over the years (Fig. 7.6).

Numerical forecasts of precipitation have been the least useful of all the outputs from numerical models. Precipitation is the end result of a long chain of atmospheric processes, with errors in each link of the chain tending to multiply in quantitative precipitation forecasting. Figure 7.7 illustrates the skill record over 19 years at the NMC for manual forecasting of accumulated precipitation greater than 25 mm during the first 24 h after issuance. The measure of skill used here is threat score, which relates the area of such precipitation correctly predicted to the total area of precipitation either observed or predicted. There is little, if any, evidence for improvement in skill since the early 1960s. A major reason for the relatively low level of forecasting skill is the problem of predicting the occurrence and distribution of summer mesoscale rain systems such as thunderstorms. This is readily apparent in an examination of seasonal variations in quantitative precipitation forecasting skill, which show much lower levels of skill during the summer months (Fawcett, 1977).

The assessment of long-range forecasts has received some attention in Britain, with some concern about the methods of assessment themselves. Jolliffe and Foord (1975) constructed a measure of association between forecast weather and observed weather. Data for monthly forecasts of weather over the UK issued by the Meteorological Office were compared with weather observations published in *Weather Log* put out by the Royal Meteorological Society. The measure of association was computed for temperature, rainfall and sunshine. The values obtained were found not to differ significantly from zero,

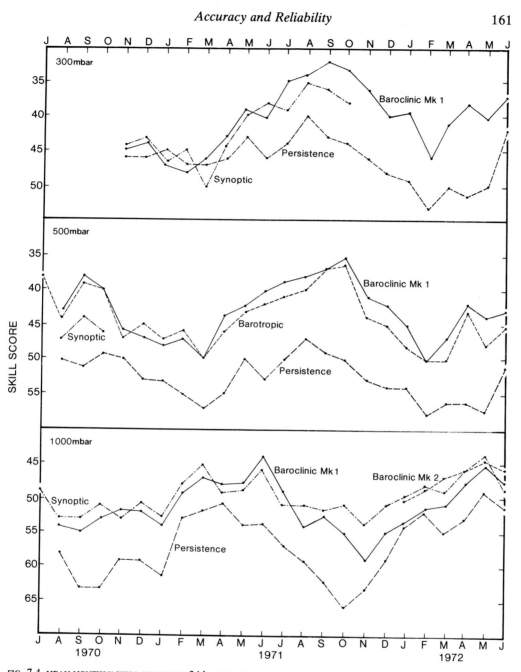

FIG. 7.4. MEAN MONTHLY SKILL SCORES OF 24 h OPERATIONAL PREDICTIONS OF BROAD-SCALE WIND PATTERN OVER THE AUSTRALIAN REGION. A LOWER SKILL SCORE INDICATES A MORE ACCURATE PREDICTION
(After Gibbs, 1972; reproduced with the permission of the Australian Bureau of Meteorology, Department of Science and the Environment).

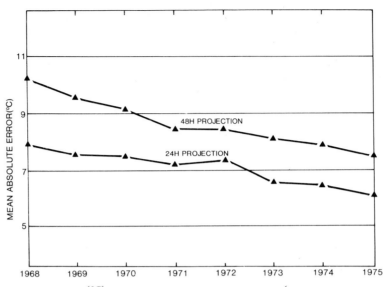

FIG. 7.5. MEAN ABSOLUTE ERRORS (°C) OF AUTOMATED TEMPERATURE FORECASTS (COMBINED MAXIMUM AND MINIMUM) FOR 126 CITIES, 1968 TO 1975
(After Fawcett, 1977; reproduced from the *Bulletin of the American Meteorological Society*, p. 145, with the permission of the American Meteorological Society).

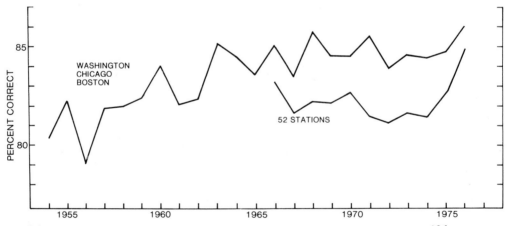

FIG. 7.6. RECORD OF SKILL OF PUBLIC FORECASTS OF WHETHER OR NOT PRECIPITATION WILL OCCUR IN A 12 h PERIOD. THE MEASURE OF SKILL IS PER CENT CORRECT
(After Shuman, 1978; reproduced from the *Bulletin of the American Meteorological Society*, p. 13, with the permission of the American Meteorological Society).

the average value which would be obtained from a set of random forecasts. On the basis of these results Jolliffe and Foord were understandably sceptical about the success of long-range forecasts. On the other hand, earlier work by Wright and Flood (1973) gave a more optimistic view of the forecasts, perhaps because of the different approach incorporating 'general information' as well as individual variables.

Gordon (1974) devised a classification scheme for evaluating the accuracy of temperature and rainfall predictions for the London area. He also found that the results of long-range forecasts for a month ahead did not approach an acceptable level of statistical significance and was reluctant to claim that they were even marginally better than chance.

Following the work by Jolliffe and Foord (1975) and Gordon (1974), Jenkinson (1975) attempted a further assessment of the accuracy of the Meteorological Office long-range weather forecasts. Instead of considering separately one or more of the ten climatic regions for which monthly forecasts were issued, Jenkinson presented an overall comparison of forecast and actual values of temperature and rainfall for the ten regions combined and for all forecasts issued. The results of the analysis clearly suggested that both temperature and rainfall forecasts had significant merit, and application of the Jolliffe and Foord measure of association also gave better results than when used by the latter.

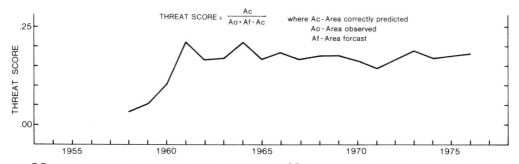

THREAT SCORE $= \dfrac{Ac}{Ao \cdot Af - Ac}$ where Ac - Area correctly predicted
Ao - Area observed
Af - Area forcast

FIG. 7.7 RECORD OF SKILL OF THE NMC GUIDANCE FOR MORE THAN 25 mm OF ACCUMULATED PRECIPITATION DURING THE FIRST 24 h AFTER ISSUANCE. THE MEASURE OF SKILL IS THREAT SCORE
(After Shuman, 1978; reproduced from the *Bulletin of the American Meteorological Society*, p. 13, with the permission of the American Meteorological Society).

Further discussion focusing on the suitability of the assessment methods has come from Wright (1976), who considered that the difference between Jenkinson's results and those of Jolliffe and Foord is unlikely to be attributable to the use of different category boundaries. The main reason is seen to be that different sets of forecasts were assessed. To confuse the issue further Wright expresses doubt about the validity of some of Jenkinson's methods and conclusions.

The overall picture emerging from these sometimes conflicting and confusing assessments of the accuracy of weather forecasts is that for periods up to 48 h weather forecasts of considerable skill and utility can be obtained. Nevertheless, small-scale features embedded in the major weather systems can cause variations that are difficult to predict, particularly in areas with irregular topography. The exact location of some highly significant weather phenomena, such as tornadoes and severe thunderstorms, cannot be forecast accurately with any skill beyond a few hours, although it is sometimes possible to predict the general area of severe storm activity up to 24 h ahead. For periods up to five days moderately skilful daily temperature forecasts are possible, but precipitation forecasts are much less reliable. For periods up to a month ahead there is some skill in predictions of average temperature conditions, but forecasts for individual days or even weeks within the period have shown little skill. There is some skill in prediction of total precipitation amounts for periods of five to seven days ahead. There may be some skill in some forecasts for periods of more than one month ahead, but skill in day-to-day forecasts is non-existent and only minimal in seasonal outlooks and climate forecasts (Amer. Met. Soc., 1976).

7.4 DISSEMINATION AND APPLICATION OF WEATHER FORECASTS

Allowing for the fact that weather forecasts are not necessarily as accurate as consumers would like, there is nevertheless a vast market for such forecasts. Consumers, actual and potential, include the general public, all branches of the transport industry, commercial enterprises from small businesses to major stores, service industries, the recreation and tourist industries, safety organizations, construction enterprises, agriculture, forestry, consulting firms and manufacturing industries, encompassing a whole range of decision makers through to the highest levels in companies, research institutions and government departments.

Several factors, as well as accuracy, may influence the usefulness of forecasts. These include the nature and speed of forecast dissemination, the extent to which forecasts are presented in terms readily understood by and relevant to the consumers, and the type of application envisaged for the forecast. McIntyre (1970) and Taylor (1971) both saw the problem as essentially one of communication and marketing, an observation which is still valid.

McIntyre identified four steps in the marketing function: packaging, presentation of the package, interpretation, and presentation of interpretation. In the packaging step meteorological information of value to the consumer is put together in appropriate form, such as the written forecasts prepared for the public or for commercial aviation. Presentation of the package involves transfer of the packaged information to the interpreter, perhaps using displays of forecast information or briefing sessions. At the interpretation stage the packaged information received through the presentation operation is interpreted in terms of the user's actual needs. This may include blending with other special meteorological or non-meteorological information and therefore requires an understanding of meteorology as well as of the business of the user. The television weathercaster is the best example in public forecasting. The final step of presentation of the interpretation is when the user gets the information he wants.

Information received at one step may produce a demand for other information. Feedback is important as an indicator of customer reaction, perhaps leading to improved forecasts for specific purposes.

Speed of transmission of forecasts is a critical element in the whole procedure. There is now a large variety of methods and devices available for the rapid and efficient communication of weather forecasts. Newspapers, radio and television, particularly the latter are important for transmitting the product to the consumer, especially the general public. Television presentations of weather forecasts, although restricted by time constraints, now make regular use of weather charts, radar images and satellite pictures to help to convey messages. Unfortunately, there is still some doubt whether many of the recipients of such forecasts really fully understand the meaning of terms such as 'cold front' or 'anticyclone' or 'blocking' or 'upper cold pool'. In addition, while it is now possible to point to some of the features themselves on satellite pictures, it is arguable whether the majority of viewers really know what they are seeing. Several television weathercasters have now produced their own little books as aids to understanding their forecasts. In some cases this may be venturing on to dangerous ground by creating grossly oversimplified versions of the complexities of the atmosphere and so leading the layman into wondering why weather forecasters seem to have so much trouble in producing accurate forecasts!

An interesting study by Beebe (1970) compared the ratings of television weather programmes presented by professional members of the American Meteorological Society with those presented by non-professional personnel. The public clearly expressed a strong preference for those stations using professional meteorologists. It appeared that the cost to either stations or sponsors was very small relative to the increased numbers of viewers of programmes presented by professionals, suggesting that meteorologists should take a more active part in promoting their services to broadcasters.

Facsimile and closed circuit television can be used to transmit weather information and forecasts to more specialized consumer groups. It is also possible to envisage greater use of FM radio channels, perhaps reserved for meteorological information, and of telecommunications satellites.

One of the major problems in weather forecasting is the misinterpretation of the forecast by the general public. Several surveys have been conducted concerning the public understanding of forecast terminology (e.g. Rogell, 1972). Many descriptive phrases used in forecasts are either poorly understood or not understood at all, and mean different things to different people. The survey results have shown that the public frequently misinterprets precipitation probability forecasts. Murphy (1977a) has suggested that this lack of understanding relates to event misinterpretation more than to misinterpretation of the probability as such. Thus, most people do not understand the definition of the event in question. Murphy sees the need for improved forms of survey, and a need to educate the public regarding the proper interpretation and use of forecasts. At the same time forecasters themselves could greatly benefit from a training programme related to the formulation, interpretation and evaluation of probabilistic forecasts, since many forecasters when formulating forecasts seem to have in mind a definition of the precipitation event which differs from the official definition (Murphy and Winkler, 1974; Murphy, 1978).

The ultimate tests for a weather forecast are in terms of its usefulness for a specific purpose and its economic returns. Throughout the book so far there have been many references to the importance of weather and weather knowledge to a wide range of activities, most of which can benefit in some way from the use of weather forecasts. A few examples will be mentioned here to give some impression of this range.

Property damage associated with severe downslope windstorms in Boulder, Colorado, averages about US$ 1 million every year (Bergen and Murphy, 1978). A study of the potential economic and societal impacts of improved short-term forecasts in the Boulder area revealed that significant increases would occur in the use of a variety of protective actions if 80 per cent accurate windstorm forecasts were available (Table 7.1). It was estimated that accurate forecasts could reduce residential property damage by about US$ 200 000 annually, with potential savings to local businesses put at an additional

Table 7.1

SAMPLE[a] OF RESPONSES FROM PERMANENT HOME RESIDENTS (PHR), MOBILE HOME RESIDENTS (MHR), BUSINESSES (LB), AND CONSTRUCTION COMPANIES (LCC) IN BOULDER, COLORADO, TO SPECIFIC PROTECTIVE ACTIONS IF 80 PER CENT ACCURATE WINDSTORM FORECASTS WERE AVAILABLE

(after Bergen and Murphy, 1978)

Protective Action	% of sample currently taking action with existing forecasts				% of sample who would take action with accurate forecasts			
	PHR	MHR	LB	LCC	PHR	MHR	LB	LCC
Close draperies	62·5	59·3			71·0	65·8		
Secure loose objects outside	53·6	65·8	26·5	88·5	72·3	72·3	33·8	88·5
Close shutters	57·2	83·7			85·7	100·0		
Move vehicles to protected area	51·8	32·0	53·8	50·0	67·8	41·3	82·1	65·4
Tape windows	9·6	10·0	10·3		32·5	21·5	20·6	
Cover car with tarpaulin	0·7	6·3			7·7	21·5		
Cancel planned shopping trips or visits	36·7	55·7			69·3	72·2		
Brace fences	8·6	4·8			16·5	14·3		
Reinforce incomplete construction	9·0			77·0	19·3			84·6
Leave work early	6·6	21·5			24·7	29·2		
Not go to work	7·2	17·7			20·5	27·9		
Not send children to school	21·4	77·0			61·8	77·0		
Pick up children from school	26·2	84·6			57·2	92·3		
Leave area for short period	1·2	17·7			4·2	29·1		
Leave area, go to motel	0·0	2·5			2·4	6·3		
Secure or remove awnings	18·8	13·3			31·2	20·0		
Cover windows with temporary protection	5·4	11·4	10·3	23·0	30·1	22·8	16·2	65·4
Increase tie-downs on mobile homes	1·2	3·8			4·2	10·0		
Disconnect gas lines		3·1				9·2		
Check tie-downs		14·5				23·2		
Secure steps to mobile homes		13·9				27·9		
Place rocks or tyres on top of home		8·9				12·7		
Lock west-facing doors		47·1				58·8		
Dismiss employees early		20·6	15·4			27·9	42·3	
Close business early		16·2				25·0		
Not open business		4·4				10·3		
Stop construction			26·9				50·0	
Reschedule daily work assignments			15·4				42·3	
Dismantle incomplete construction			3·8				15·4	

[a] sample defined as the number of respondents who have the item, when applicable

US$ 150 000 (Table 7.2). Such benefits appear to greatly exceed any incremental costs associated with formulation and dissemination of the forecasts.

On a larger scale, Anderson and Burnham (1973) demonstrated that for residential and retail activities in the US Gulf of Mexico coastal region a reduction of average hurricane

Table 7.2
WINDSTORM LOSSES AND PREVENTABLE DAMAGE ESTIMATES FROM LOCAL BUSINESSES
AND CONSTRUCTION COMPANIES IN BOULDER, COLORADO, FOR THE WINTERS OF
1971–72, 1972–73 AND 1973–74
(after Bergen and Murphy, 1978)

| Enterprise | Winter | Losses (US$) | | | Preventable Losses (US$)[a] | | | |
		Business Sales / Construction Labour	Materials	Total	Sales / Labour	Materials	Total	%
Local businesses	1971–72	57 305	78 100	135 405	2 000	43 650	45 650	33·8
	1972–73	27 455	63 325	90 780	2 000	42 250	44 250	48·8
	1973–74	22 505	74 150	96 655	12 000	53 180	65 180	67·2
Local construction companies	1971–72	31 700	32 799	64 499	7 050	5 950	13 000	20·2
	1972–73	18 000	32 874	50 874	3 600	3 525	7 125	14·0
	1973–74	8 350	5 655	14 000	850	510	1 360	9·7

[a] assuming availability of 80 per cent accurate forecasts with 2–4 h lead times

forecast error by 50 per cent would save at least US$ 15·2 million in the first year, for the 20 per cent of the population already protecting. Assuming that a growing proportion of the population adopts protective measures, partly encouraged by improved forecasts, the savings from such forecasts will also rise (Table 7.3).

Table 7.3
BENEFITS TO THE US GULF OF MEXICO REGION RESULTING FROM A 50 PER CENT
REDUCTION IN HURRICANE FORECAST ERRORS
(after Anderson and Burnham, 1973)

Year	Proportion that protects (%)	Savings from improved forecasts (US$ millions)
1	28	21·489
2	36	27·998
3	68	57·380
4	100	88·437
	Total Savings	195·437

The potential benefits from improvements in tropical cyclone forecasts in the western North Pacific to the US Department of Defense, have been examined by Brand and Blelloch (1975). Tropical cyclone forecasts influence the decisions of military commanders responsible for about fifty ports, ten major Air Force bases and numerous other military installations in the tropical cyclone area of the western North Pacific. Even a single wrong decision may cost as much as US$ 10 million, and improved forecasts are shown to carry potential savings of many millions of dollars each year.

Glantz (1977), in a penetrating study of the value of a long-range weather forecast for the West African Sahel, makes the point that such a forecast may not even be desirable until there have been some essential adjustments to existing social, political and economic practices. The theoretical advantages to the Sahelian governments of a reduced precipitation forecast six months ahead seem clear enough: previously developed programmes could be activated to reduce the impact. Unfortunately, the various political, social and economic constraints are seen to be so strong as to almost completely eliminate any possible benefits to be derived from any long-range weather forecasts, however accurate.

In any particular situation the value of weather forecasts depends upon many factors, including the nature of the forecasts themselves. Murphy (1977b) identified three principal types of weather forecast currently produced on a routine operational basis as: climatological forecasts, categorical or deterministic forecasts, and probabilistic forecasts. He examined the value of each type in the cost−loss ratio situation, which involves a decision maker who must decide whether or not to take protective action with respect to some activity or operation, in the face of uncertainty as to whether or not the forecast is accurate. The most important implication arising from his results is that the value of even moderately unreliable probabilistic forecasts generally exceeds the value of categorical and climatological forecasts. This suggests that the value of day-to-day forecasts could be significantly increased if such forecasts were routinely expressed in probabilistic terms and disseminated more widely to decision makers. It is important to note that the benefits which might accrue do not depend on scientific advances in weather forecasting, rather they depend on appropriate education of the forecasters and the decision makers alike.

7.5 THE FUTURE

There are as many opinions on the future of weather forecasting as there are approaches to the forecasting problem. Optimism seems to predominate, but some meteorologists, such as Ramage (1978), see the need for new directions in the organization of weather forecasting services, or for marked deviations in traditional thinking (Beran et al., 1977).

Ramage suggests that numerical weather forecasting skill has neither improved nor is likely to improve measurably, and that better forecasts will only come from earlier detection of and faster response to important events, plus more careful tailoring of forecasts to fit needs. He has proposed a major restructuring of the USA public weather service to make better use of local skills and talents, to ensure local accountability and to reflect local priorities. This would involve close collaboration between federal weather agencies such as the National Weather Service and the National Climatic Center, and universities and industrial meteorologists at the local level. It is envisaged that university departments of meteorology, combining their existing teaching, research and service functions, could look after local forecasting, with the expanded role of industrial meteorologists to provide specialized forecasts.

A much more optimistic view has been offered by Reed (1977) in anticipating a bright future for weather prediction, with numerical modelling in a healthy state of development, its full potential still to be realized. Shuman (1978) supports the argument that the skill of numerical weather prediction can be increased and the period of operational predictions extended. Increased model resolution, perhaps beyond the capabilities of present computers, may lead to increases in forecast accuracy, but this will also require advances in model physics and in input data density.

Particularly important problems remain to be solved in the application of numerical techniques to the forecasting of smaller-scale weather phenomena such as thunderstorms and heavy precipitation (Fawcett, 1977). Some preliminary results predicting hurricane motion and severe convective systems have been promising, offering considerable hope for the future, but many models of severe weather will have restricted operational utility because of their complexity and expense (Pearson, 1976b). Although there seems to have been little improvement in quantitative precipitation forecasts since the early 1960s, Shuman (1978) suggests that recent research results point to the greatest improvements in this area during the 1980s.

There are limits to the predictability of the atmosphere. Theoretical work on atmospheric predictability has indicated that the intrinsic properties of the atmosphere, together with the problems of observing its finest-grain details of behaviour, impose an upper limit on the prediction of day-to-day weather changes. This limit is believed to be no more than two weeks (American Meteorological Society, 1976a). Present predictive capability of about five days for current operational models is still well short of the theoretical limit, leaving considerable scope for improvement.

Among the measures which might lead to such improvement are the large international observational programmes under the auspices of the World Meteorological Organization (WMO) and the International Council of Scientific Unions (ICSU). Specifically, the combined World Weather Watch (WWW) and Global Atmospheric Research Programme (GARP) will provide improved global observing capabilities and the development of enhanced modelling through better treatment of physical processes, and improved numerical techniques and methods of data assimilation. Both WWW and GARP place heavy reliance on meteorological satellite technology, the potential of which is still largely unexploited. Other technological advances such as dual Doppler radar and acoustic sounders add to the prospects for improved forecasting capabilities.

In some areas, such as airport weather forecasting, recent developments in meteorological sensors have been rapid and have only widened the gap between observational and forecast ability (Beran et al., 1977). A need is seen for new approaches to narrow the technological gap between weather forecasting and aircraft design and utilization. Beran et al. suggest that useful models for parameterizing the planetary boundary layer are at least ten years away. They predict that while weather modification and aircraft design will undoubtedly be important, the most immediate improvements will come from departures from traditional approaches to forecasting. The change may result from a more intense effort to understand the mesoscale, especially the anomalies, coupled with the use of radically new observational techniques, particularly radar for measuring wind fields in clear air, and perhaps radiometric and optical techniques.

The future for weather forecasting is at best hazy. Improvements will almost certainly occur but it is difficult to determine their exact nature. There is no doubt that technology will improve, but there is some doubt that data networks and data handling techniques will keep pace. Growth can be envisaged in the contributions made by professional consulting meteorologists, particularly in Europe, accompanied by better application of local knowledge to more specialized forecasts. A clear need exists for better education of users, producers and disseminators of weather forecasts, so that users know how to make best use of the information available, producers give the consumers the products they need, and the appropriate information is accurately and speedily communicated to those who want it.

ACKERMAN, W. C., 1976, 'Forecasting yes, management maybe.' *Bull. Amer. Met. Soc.*, 57, pp. 984–7.

ANDERSON, R. K., 1978, 'Current uses of satellite data in a meteorological forecast office.' *WMO Bull.*, 27, pp. 91–103.

ATLAS, D., 1976, 'Overview: the prediction, detection and warning of severe storms.' *Bull. Amer. Met. Soc.*, 57, pp. 398–411.

BROMLEY, E., 1975, 'Aviation weather forecasts in tomorrow's flight service system.' *Bull. Amer. Met. Soc.*, 56, pp. 372–4.

GLANTZ, M. H., 1977, 'The social value of a reliable long-range weather forecast.' *Ekistics*, 43, pp. 305–13.

KNIGHTING, E., 1976, 'The weather a week ahead.' *Nature*, 262, pp. 162–3.

LANSFORD, H., 1975, 'Weather watchers.' *Nature*, 256, pp. 688–90.

LAURMANN, J. A., 1976, 'Variance estimates in seasonal climate forecasting and related food reserve requirements.' *J. Appl. Met.*, 15, pp. 529–34.

LIVEZEY, R. E. and JAMISON, S. W., 1977, 'A skill analysis of Soviet seasonal weather forecasts.' *Mon. Weath. Rev.*, 105, pp. 1491–500.

McPHERSON, R. D., 1975, 'Progress, problems and prospects in meteorological data assimilation.' *Bull. Amer. Met. Soc..*, 56, pp. 1154–66.

MOORMAN, T., 1968, 'The forecasting problem.' *Bull. Amer. Met. Soc.*, 49, pp. 12–15.

MORGAN, J., 1978, 'Meteosat-1 in orbit.' *WMO Bull.*, 27, pp. 250–3.

MUENCH, H. S., 1976, 'Use of digital radar data in severe weather forecasting.' *Bull. Amer. Met. Soc.*, 57, pp. 298–303.

MURRAY, R., 1970, 'Recent developments in long-range forecasting in the Meteorological Office.' *Quart. J. Roy. Met. Soc.*, 96, pp. 329–36.

NAMIAS, J., 1968, 'Long-range weather forecasting: history, current status and outlook.' *Bull. Amer. Met. Soc.*, 49, pp. 438–70.

PIELKE, R. A., 1977, 'An overview of recent work in weather forecasting and suggestions for future work.' *Bull. Amer. Met. Soc.*, 58, pp. 506–20.

RATCLIFFE, R., 1970, 'Sea temperature anomalies and long-range forecasting.' *Quart. J. Roy. Met. Soc.*, 96, pp. 337–8.

SCOGGINS, J. R. and VAUGHAN, W. W., 1971, 'How some nonmeteorological professionals view meteorology and weather forecasting.' *Bull. Amer. Met. Soc.*, 52, pp. 974–9.

SOMERVILLE, R. C. J., 1977, 'Pattern recognition techniques for forecast verification.' *Beiträge zur Physik der Atmosphäre*, 50, pp. 403–10.

THOMPSON, P. D., 1977, 'How to improve accuracy by combining independent forecasts.' *Mon. Weath. Rev.*, 105, pp. 228–9.

TRENBERTH, K. E. and NEALE, A. A., 1977, 'Numerical weather prediction in New Zealand.' *Mon. Weath. Rev.*, 105, pp. 817–25.

VAN LOON, H. and JENNE, R., 1975, 'Estimates of seasonal mean temperature using persistence between seasons.' *Mon. Weath. Rev.*, 103, pp. 1121–8.

WILLIAMS, J., 1977, *Can we predict climate fluctuations?* Professional Paper, PP-77-7, (International Institute for Applied Systems Analysis, Laxenburg, Austria).

WINKLER, R. L. and MURPHY, A. H., 1976, 'Point and area precipitation probability forecasts: some experimental results.' *Mon. Weath. Rev.*, 104, pp. 86–95.

8
Management of Atmospheric Resources

In this chapter there is a continued examination of the ways in which man attempts to manage his atmospheric resources, and of the use which is or might be made of the expertise of meteorologists and climatologists. Management strategies for the atmosphere range from direct attempts to modify weather and climate to developments in international programmes of observation and research.

8.1 WEATHER AND CLIMATE MODIFICATION

There has been a great deal of research into, and much has been written about, weather modification. The very broad field is covered comprehensively in several books (see particularly, Hess, 1974 and NAS, 1973) and in a multitude of journal articles. This review can do little more than present a very brief summary of the present state and possible future developments of weather and climate modification. Weather modification is the most direct way in which man attempts to manipulate the atmospheric resource, yet it provides a good illustration of the magnitude of the management problems and it highlights the present state of imperfect understanding about the working of the atmosphere.

The history of scientific weather modification attempts starts in 1946 with the dry ice seeding experiments of Schaefer and Langmuir, but man's attempts to control the weather and improve the climate date back to ancient times. The earliest attempts to influence the weather consisted of prayers, sacrifices and symbolic dances, all of which still find adherents today. Man has constantly sought means of modifying the timing, location and magnitude of precipitation, temperature and winds, using methods such as ringing bells, beating drums, and firing rockets and cannon.

Precipitation has been the prime objective of weather modification efforts because of man's dependence on it to provide for a multiplicity of uses. Cloud seeding has been attempted for a variety of purposes such as hail suppression, fog dissipation, reduction of the force of tropical cyclones and suppression of lightning, but the main focus of investigation and experimentation has been on the augmentation of precipitation (Fig. 8.1). Early successes with the artificial nucleation of ice in supercooled clouds led to considerable optimism that seeding produces increases of precipitation, and experiments in which there appeared to be no increases, or even decreases in precipitation reaching the ground were discounted or disregarded by the enthusiasts (Neiburger, 1973).

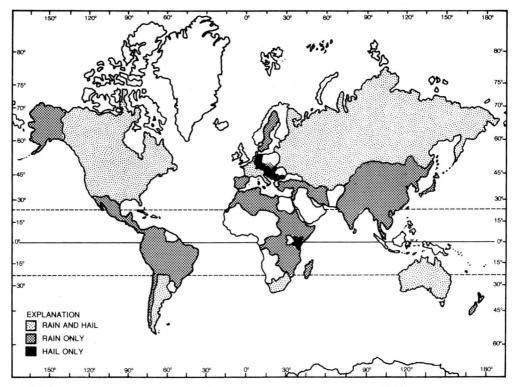

FIG. 8.1. NATIONS IN WHICH WEATHER MODIFICATION PROJECTS HAVE OCCURRED SINCE 1945
(After Changnon, 1975; reproduced from the *Bulletin of the American Meteorological Society*, p. 29, with the permission of the American Meteorological Society).

In Australia, the announcement in February 1947 of a success in scientific rainmaking through cloud seeding was, not surprisingly, greeted with great enthusiasm in a land of recurring droughts. The scientists involved, however, were well aware that clouds capable of producing rain, an important prerequisite for successful seeding, tend to be rare things during droughts. General acceptance of the thesis that cloud seeding may produce increases, decreases or no change in the amount of precipitation reaching the ground, depending on circumstances, has come gradually.

For example, cumulus cloud in maritime air already contains ample nuclei, so that artificial seeding merely increases the number of nuclei and may thus reduce precipitation. The most productive seedable clouds are orographic, forming over continental areas. Such clouds are more likely than those developing over the sea to be of the cold-topped type susceptible to seeding. Evidence suggests that rainfall from these clouds tends to be increased by being of longer duration rather than by having an increased intensity. Experiments in Tasmania's predominantly maritime air have indicated that seeding stratiform clouds may produce additional rain, while seeding cumuliform clouds will not.

Present capability to predict the impact of seeding in any area needs to be based on evidence from lengthy controlled experiments, because rainfall varies so widely in the natural course of events (Anon, 1978a). Increased understanding of how clouds work and improvements in techniques for gathering information about them should, in the not too distant future, make cloud seeding generally less of a gamble. Changnon (1975) has noted significant advances in some areas of cloud seeding to enhance precipitation. Orographic clouds can be modified to yield snow increases of up to 30 per cent under certain conditions in mountainous areas, and rainfall from single tropical cumulus clouds can be markedly enhanced (Gagin and Neumann, 1974; Grant and Kahan, 1974; Simpson and Dennis, 1974; Smith, 1974).

A major problem with the use of cloud seeding to stimulate rainfall still lies in assessing its effectiveness (Olsen and Woodley, 1975). There is no way of telling, if rain follows seeding, whether any of it was in fact produced by the rain-makers.

There has been considerable expenditure of money and effort in attempts to modify tropical cyclones and suppress hail, but there is no wide agreement as to the effectiveness of the techniques used. The basis of the hurricane modification experiments which have been carried out since 1961 under Project Stormfury in the USA is that massive ice-nucleus seeding on one side of the zone of maximum winds will release latent heat, increase vertical motion promoting the formation of convective cloud and so decrease the flow of air into the vortex and promote a corresponding decrease in the strength of the winds. The experiments are so far inconclusive but encouraging. They are continuing and may be extended to include tropical cyclones in the Australian region (Gentry, 1974).

The prevention of crop damage by hail has been the objective of weather modification attempts for many centuries. In France, Italy, Switzerland and Australia, cannon, guns and rockets have been fired into thunderstorms with apparent success to the extent that many farmers continue to believe in the methods. Field tests with dry ice or silver iodide seeding have been carried out in many countries with particular success being claimed in the USSR (Sulakvelidze et al., 1974).

Dispersal of cold fog (fog consisting of liquid drops at temperatures below 0 °C) is now a well-established procedure. Seeding is carried out by dry ice either dropped from planes or ejected from ground apparatus, or by the emission of liquid propane from dispensers on the ground. Routine cold fog dissipation by these methods is carried out at several airports in the USA, the USSR and in France, with reported effectiveness better than 80 per cent.

Warm fogs are among the most stable cloud systems in the atmosphere and therefore much harder to deal with. They are also more frequent hazards at most of the major and minor airports around the world, so considerable effort has been directed to development of dispersal methods. Three main techniques are all designed to promote evaporation of the water droplets in warm fog: mechanical mixing of the fog with drier, warmer air from above, through the use of helicopter downwash; drying of the air with hygroscopic chemicals, which tends to be costly; and heating of the air. Ground-based heating is one of the oldest and most successful methods of dissipating warm fog and is effective in all warm fog situations. It is relatively simple and inexpensive to operate and maintain but extremely costly to install (Silverman and Weinstein, 1974).

On a much larger scale there have been many schemes proposed for modifying climate, with possible effects extending beyond national boundaries perhaps even globally. Several grandiose engineering schemes have suggested projects such as diverting rivers which flow into the Arctic Ocean, damming the Bering Strait to change the ocean circulation, inducing the Gulf Stream to flow into the Arctic Ocean to melt the pack ice, creating huge artificial lakes or inland seas in the Sahara and Australia, coating parts of the Australian desert with bitumen, or building artificial mountain ranges to promote precipitation in central Australia, and damming Drake Passage. Clearly, at this sort of scale the modification attempts of one nation might adversely affect the welfare of another. The probable climatic consequences of most such proposals are not yet understood. Indeed, apart from a recognition that the effects of cloud seeding may extend at least 200 km downwind of the target area, there is little understanding of the inadvertent effects of weather and climate modification.

The purpose of most weather modification is to improve productive efficiency and reduce hazard dangers. Increased precipitation at the right time and in the right place, for example, may mean the difference between success and failure of a crop or may improve the operating efficiency of a hydroelectric power station. On the other hand, increased precipitation which benefits the wheat farmer may ruin the crop of a fruit farmer in the same area. It is also uncertain whether an increase in precipitation in one area is at the expense of another area downwind. The point is that the whole question of weather modification is one that is loaded with economic, social, legal and political implications, most of which remain to be clarified. Hess (1974) has suggested that it may even turn out to be easier to develop ways to modify weather beneficially than to develop ways to apply this knowledge.

It is not easy to produce effective legislation relating to weather modification. Davis (1974) provides an interesting discussion of weather modification litigation and statutes which raise some fascinating questions. Attempts at precipitation augmentation, for example, may pit the modifiers against environmentalists and conservationists. Social, political and legal forces are becoming more involved in decisions regarding the applications and effects of weather modification (Changnon, 1973). The public is becoming more aware of local and large-scale weather modification activities, which are usually accompanied by new controversies bringing increasing interest to legislation at the national level. The emerging capability of various nations to modify weather and climate, perhaps using modification as a weapon, has led to numerous appeals for peaceful uses of the technology and for international cooperation (Charak and DiGiulian, 1974; DiGiulian and Charak, 1974; MacDonald, 1975).

The future of weather modification is unclear. On one hand it is possible to point to the many potential economic and social benefits; on the other hand weather modification technology is still immature, the socioeconomic impacts are ill defined, management has been uncertain, and public acceptance is clouded by a series of public, scientific, legal political, and military controversies (Changnon, 1975). In the USA, where federal support of weather modification research and operations approached US\$ 20 million per year in the early 1970s, declines in funding in 1974 and 1975 may have reflected the weight of such factors. In Australia, however, in early 1979 the CSIRO was reported to be ready to invest about A\$ 1.25 million over five years trying to discover if artificial rainmaking is economically viable for farmers. If all goes to plan this largest yet rainmaking exercise in Australia could boost the value of wheat crops in western Victoria by more than A\$ 1 million in the first year.

8.2 ECONOCLIMATE

During the last decade there has been growing interest in a totally different approach to the problem of atmospheric resource management. It derives from the idea that information about the atmosphere, in the form of weather and climate data, is an important resource in itself. From this it follows that, given an understanding of man's interactions with the atmosphere and given adequate information about atmospheric events, man can use his management ability to improve the economic and social outcomes of weather-sensitive activities (Sewell, 1968; McQuigg, 1970; Maunder, 1974). This is the field econoclimate (Perry, 1971b).

There have been several attempts to develop measures of climatic effects on economic activities which are precise enough to permit the assessment of costs and benefits of programmes designed to improve the outcomes of weather-related activities (Johnson and McQuigg, 1974). Determination of associations between economic and climatic data is not easy, however, as economic information relates to areas and climatic data to places (Maunder, 1974). This necessitates some form of transformation, usually of the climatic data.

The work of Maunder is especially noteworthy in this regard. He has produced a large number of econoclimatic studies, based mainly on New Zealand examples. A good example of his methods is contained in his study of relationships between weighted rainfall, temperature and water deficit indices and New Zealand dairy production on the one hand, electric power consumption on the other (Maunder, 1972; 1974). Development of such weighted climatic indices enables a vast amount of climatological data to be conveniently expressed as an index representative of the climatic data and also meaningful to economic activities.

In the study of New Zealand dairy production the weightings applied to the three selected climatic elements were land area, human population, sheep population, dairy cow population, beef cattle population, and crop area. Then, for each month assessed, six weighted indices were computed for each of rainfall, mean temperature, and days with water deficit. The weighted indices on a county basis were then combined into regional and national indices. Application of such indices to nation-wide economic activity is complicated by the lack of suitable economic indicators. In New Zealand appropriate data were available for butterfat production and consumption of electric power. Analysis of the relationship between butterfat production (expressed in terms of departure from expected production) and water deficit indices weighted by dairy cow population gave a correlation coefficient of -0.82, significant at the 0.1 per cent level. The analysis of the relationship between monthly temperature departures from normal, weighted according to the distribution of the human population, and the random oscillation of total monthly electricity generation gave a correlation coefficient of -0.50, also significant at the 0.1 per cent level.

Maunder (e.g. 1977a, b, c, d; 1978b) has made many other studies of such relationships aimed primarily at improving the effective and economic use of weather and climate information. Johnson and McQuigg (1974) adopted slightly different approaches to the same basic question, by application of principal components analysis and linear probability models. In all cases the efforts are aimed at better understanding of the relationships between climate and economic activity and hence better and more efficient use of available resources.

8.3 INTERNATIONAL COOPERATION

An important development for modern meteorology and climatology has been the growth in international cooperation. Such growth is a critical element in the management of atmospheric resources, since it facilitates research and exchange of information. International cooperation in investigations of the atmosphere is by no means a recent development, however, and can be traced back for over 300 years (Rigby, 1965).

From the middle of the seventeenth century various scientific societies, such as the Royal Society of London and the Florentine Academy of Science, provided the basis for international exchanges. The first really effective international cooperation did not come until the late eighteenth century with the establishment by John Hemmer of a network of thirty-nine stations from Siberia to Massachusetts. In 1853 Matthew Maury secured the cooperation of ten maritime powers in ship reporting. Then, in 1873 came the formation of the International Meteorological Organization (IMO), the forerunner of the present World Meteorological Organization (WMO) which was established in 1950. The International Polar Year of 1882–83 was the first large-scale project involving international scientific collaboration, encompassing meteorological and other geophysical studies. The second International Polar Year programme in 1932–33 again concentrated on the Arctic region, but included a small programme in the Antarctic. A proposal for a third such project led to the much broader scope of the International Geophysical Year (IGY) in 1957–58.

The IGY marked a major step forward in international meteorological cooperation, embracing nearly 70 nations and various international scientific bodies. The impetus of the IGY was continued through 1959 with the International Geophysical Cooperation (IGC). The results of observations for the period 1957–59, from thousands of stations throughout the world, were readily available in uniform format. The tremendous success of the IGY-IGC programme spawned later major projects such as the International Indian Ocean Expedition (IIOE) involving 27 countries over a period from 1960 to 1965, and the International Years of the Quiet Sun (IQSY) in 1964–65 with 71 countries participating.

The World Meteorological Organization, based in Geneva, has played major roles in these and subsequent programmes, particularly as a medium for the collection, processing and dissemination of data. The scope of the WMO is best summarized in terms of the stated aims behind its creation:

1. To facilitate international cooperation in the establishment of networks of stations and centres to provide meteorological and hydrological services and observations
2. To promote the establishment and maintenance of systems for the rapid exchange of meteorological and related information
3. To promote standardization of meteorological and related observations and ensure the uniform publication of observations and statistics
4. To further the application of meteorology to aviation, shipping, water problems, agriculture and other human activities

5. To promote activities in operational hydrology and to further close cooperation between meteorological and hydrological services

6. To encourage research and training in meteorology and, as appropriate, in related fields.

The member nations of the WMO are grouped into six regional associations: Africa, Asia, South America, North and Central America, South-West Pacific, and Europe. Each regional association has the task of coordinating meteorological and related activities within its area. In addition there are eight technical commissions, composed of experts with responsibility for studying meteorological and hydrological operational systems, applications and research. The technical commissions are for aeronautical meteorology, agricultural meteorology, atmospheric sciences, basic systems, hydrology, instruments and methods of observation, marine meteorology, and special applications of meteorology and climatology.

The WMO assumes responsibility for a multitude of activities including in particular cooperation programmes and training of meteorological personnel in developing countries. Information about these and the whole range of WMO activities is contained in the quarterly WMO Bulletin. Most issues of this contain details of programmes in countries such as Botswana (establishment and operation of a forecasting office), Gabon (training meteorological personnel), Haiti (utilization of wind energy for irrigation and power supply), India (flood-forecasting system), and Mexico (agrometeorological investigations). In the past the WMO has helped developing countries with technical assistance, mainly in the areas of training, telecommunications organization, improvement of observational networks, agrometeorology and hydrometeorology (e.g. Tewungwa, 1973).

The basic part of the WMO weather monitoring systems is the World Weather Watch (WWW) programme, which was first conceived in 1961 (Fig. 8.2). It has been set up in such a way that any country can receive meteorological information on any scale, including global, for making forecasts and to use in other ways, depending on requirements. The WWW comprises three global scale elements: an observing system, a telecommunication system, and a data-processing system (see Fig. 7.1.).

The observing system ensures that meteorological observations are made every few hours, usually every three or six hours, at fixed international times at a network of stations around the globe. Observations are obtained from about 10 000 land stations, several thousand ships at sea, special ocean weather ships, a large number of aircraft, and a network of meteorological satellites.

All these observations have to be exchanged within a few hours on regional, hemispheric and even global scales, if they are to be of value for synoptic forecasting. This necessitates the maintenance of a complex global telecommunication system. In addition, many centres have to be equipped with modern data-processing equipment to make use of the vast quantities of data available (Bugaev, 1973).

The WMO also maintains a programme on the Interaction of Man and his Environment, aimed at applying meteorological knowledge to various human activities. The programme employs two types of monitoring networks. Basic environment data networks gather data for long-range application. They include, for example, the basic climatological network gathering observations for elements such as temperature, pressure, precipitation,

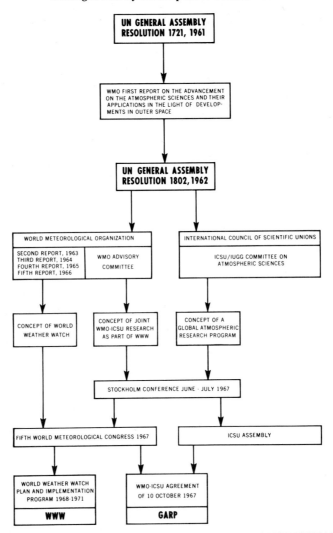

FIG. 8.2 ORIGINS OF THE WORLD WEATHER WATCH AND THE GLOBAL ATMOSPHERIC RESEARCH PROGRAMME
(After Zillman, 1977; reproduced with the permission of the Australian Bureau of Meteorology, Department of Science and the Environment).

humidity, wind and evaporation from over 100 000 stations; the network for isotope concentration in precipitation; and the global background air pollution network. In the latter network, regional air pollution stations are intended to reveal long-term changes in atmospheric composition due to regional sources, and baseline stations document long-term changes of particular significance to weather or climate on the global scale.

Such basic data networks are supplemented by various other networks designed to meet the requirements of specific activities. The network monitoring in support of marine activities is based on several thousand voluntary observing ships, together with a few stationary meteorological ships, automatic buoys and weather satellites. The planning of the network monitoring in support of air transport is coordinated jointly by the WMO and the International Civil Aviation Organization. This monitoring system assists the preparation of forecasts and statistical information needed to support aviation activities. Other networks monitor specifically in support of hydrology and agriculture (Wallen, 1974).

Two recent programmes which provide good illustrations of the continuing efforts of the WMO are the Precipitation Enhancement Project (PEP) and the World Climate Programme (WCP). The PEP is an internationally planned, executed and evaluated experiment in artificial augmentation of precipitation. The main objective is to establish the seedability of clouds in a selected experimental area in Spain. If the project is successful it should help to answer major questions about cloud seeding techniques, such as where to introduce the seeding reagent, by what method, for how long and in what concentrations. Better understanding of these problems could allow the most efficient seeding strategy to be applied under given conditions, thereby greatly improving the chances of success in precipitation enhancement projects (List, 1978).

The WCP has been devised as a result of proposals that the WMO should undertake a study of the problems of climatic change and variability, of the impact of human activities in changing the climate and of the effects of climatic changes on the human race. The overall programme is envisaged as having three major components. The climate data and applications programme will have two distinct objectives: the provision of climate data for both applications and research; and the application of the knowledge of climate to the immediate problems of society.

The second major component will be a programme for the study of impacts of climate on human activities. Studies under this programme will be designed to gain a better understanding of the role of climate in relation to environmental and socio-economic systems. It will require collaboration between the WMO and other international organizations such as FAO, UNEP, UNESCO and ICSU, and with research institutions such as the International Institute for Applied Systems Analysis (IIASA).

A third component will be a programme for research on climatic change and variability, with a primary objective to improve our knowledge of the observed structure and variations of global and regional climate and the mechanisms which govern them, and to develop models capable of simulating atmospheric behaviour. The second objective will be to facilitate the application of these models in assessing the predictability of climatic behaviour and the sensitivity of climate to external influences (Anon., 1978b, c).

Perhaps the best known of all the WMO projects is the Global Atmospheric Research Programme (GARP), initiated in 1967 and sponsored jointly by the International Council of Scientific Unions (ICSU) (see Fig. 8.2). As subsequently defined by the WMO and ICSU, GARP is a programme for studying those physical processes in the troposphere and stratosphere that are essential for an understanding of:

1. The transient behaviour of the atmosphere as manifested in the large-scale fluctuations which control changes of the weather; this would lead to increasing the accuracy of forecasting over periods from one day to several weeks
2. The factors that determine the statistical properties of the general circulation of the atmosphere which would lead to a better understanding of the physical basis of climate. This programme consists of two distinct parts, which are, however, closely interrelated:

(a) The design and testing by computational methods of a series of theoretical models of relevant aspects of the atmosphere's behaviour to permit an increasingly precise description of the significant physical processes and their interactions

(b) Observational and experimental studies of the atmosphere to provide the data required for the design of such theoretical models and the testing of their validity.

A concept central to GARP from the outset has been a global observational experiment. This was initially conceived as the First GARP Global Experiment (FGGE), but the scale of the operation involved makes it very unlikely that another could be mounted for many years. The Global Weather Experiment (GWE), as the undertaking is now named, is a vast project in the fields of science and international affairs. From December 1978 to December 1979 many thousands of scientists around the world have carried out the most intensive study and surveillance of the atmosphere and sea surface yet attempted. In addition to the routine surface-based network of weather stations, data were gathered by a variety of special observing systems. These include 5 geostationary satellites observing clouds and winds, and relaying data from other sources; several other satellites observing temperatures, humidities, ozone levels and surface winds; 300 instrumented balloons drifting around the tropics at a height of 14 km, relaying information on air temperature, and pressure; instrumented packages known as 'dropsondes' dropped from special aircraft to observe the monsoons; 300 instrumented buoys drifting in the southern oceans; about 80 commercial aircraft; and at least 40 ships, which observed the monsoons. In addition to the routine observations from such sources throughout the year of the project, there were two separate two-month periods of even more intense observations of the tropics and the southern hemisphere.

Planning for the GWE has opened up new frontiers in international cooperation in science, between nations, between disciplines, and between the governmental and non-governmental areas of science. The direct cost of the whole operation is estimated to be as high as several hundreds of millions of dollars, and indirect costs could be ten times greater. Several comprehensive accounts of the GWE have been published. The interested reader should consult them for further details (e.g. Dyer, 1975; Nkemdirim, 1975; Anon., 1976b; Zillman, 1977; Anon., 1978d).

As can be seen from Figure 8.3, there is a great deal more to GARP than the grand project of the GWE. Many sub-programmes have been undertaken, ranging from the GARP Atlantic Tropical Experiment (GATE) in 1974, to the mountain subprogramme (ALPEX) planned for 1980. GATE was the first major observational experiment in GARP, aimed at extending knowledge of the meteorology of the equatorial zone, with a view to improving understanding of the atmospheric circulation as a whole and prediction in the tropics in particular. Much has been written about GATE (e.g. Julian and Steinberg, 1975; Mason, 1975; Kuettner and Parker, 1976; Simpson, 1976; Perry and Walker, 1977).

International initiatives in meteorology have not been confined to the WMO. In October 1973, after four years of preliminary work, a convention was signed establishing the European Centre for Medium Range Weather Forecasts. The signatories were Belgium, Denmark, the Federal Republic of Germany, Finland, France, Greece, Ireland, Italy, the Netherlands, Portugal, Spain, Sweden, Switzerland, the UK, and Yugoslavia.

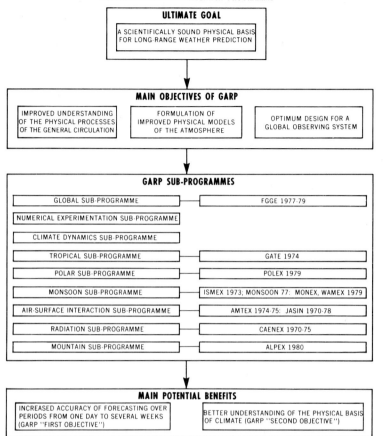

GLOBAL ATMOSPHERIC RESEARCH PROGRAMME

FIG. 8.3. A SUMMARY OF THE ULTIMATE GOAL, OBJECTIVES, SUB-PROGRAMMES AND POTENTIAL BENEFITS OF GARP (After Zillman, 1977; reproduced with the permission of the Australian Bureau of Meteorology, Department of Science and the Environment).

The convention has since been signed by Austria and Turkey. The centre, which is situated at Shinfield Park, Reading, England, is still in the development stages, with expectations that the first operational forecast will be made in 1980 (Knighting, 1978). The objectives of the centre, laid out in the convention, are summarized in Table 8.1. The benefit to Europe of reliable weather forecasts for a week or so ahead has been estimated at more than £ 100 million at 1970 values, so the centre should play an important role in the economic well-being of the member countries (Knighting, 1976).

Developments such as this, together with the establishment in the USA of a Center for Climatic and Environmental Assessment (since expanded to become the Center for Environmental Assessment Services with a subsidiary Climatic Impact Assessment Division) and of a similar 'Special Services' unit in the New Zealand Meteorological Service, have prompted a proposal for the establishment of an Australian and New Zealand Centre for Econoclimatic Assessments (Maunder, 1979). Initially this would be concerned with

Table 8.1
OBJECTIVES OF THE EUROPEAN CENTRE FOR MEDIUM RANGE FORECASTS
(after Knighting, 1978)

a. Development of dynamic models of the atmosphere with a view to preparing medium range weather forecasts by means of numerical methods;

b. Preparation, on a regular basis, of the data necessary for the preparation of medium range weather forecasts;

c. To carry out scientific and technical research directed towards the improvement of these forecasts;

d. Collection and storage of appropriate meteorological data;

e. To make available to the meteorological offices of the Member States, in the most appropriate form, the results of the studies and research provided for in the first and third objectives above and the data referred to in the second and fourth objectives;

f. To make available a sufficient proportion of its computing capacity to the meteorological offices of the Member States for their research, priority being given to the field of numerical forecasting. The allocation of the proportions would be determined by Council.

g. To assist in implementing the programmes of the World Meteorological Organization;

h. To assist in advanced training for the scientific staff of the meteorological offices of the Member States in the field of numerical weather forecasting.

monitoring on a global scale the effect of weather and climate on pastoral production. The implications of such monitoring, and perhaps predictions, are clearly of major political, economic and perhaps even military significance.

8.4 ROLE OF THE METEOROLOGIST

Growing public awareness of the environment in which we live and increasing concern for the finite atmospheric resource have created new demands on the science of meteorology. This has given to the meteorologist a role of major importance to the future quality of life (Tucker, 1976). Many of the possible contributions from meteorologists and climatologists have already been identified in earlier sections of this book, and the concern here is to highlight areas in which future contributions might prove most significant. The work of the WMO Technical Commission for Special Applications of Meteorology and Climatology (CoSAMC) provides a useful guide. The commission has been particularly active in meteorological work related to increasing energy shortages. Since energy consumption for both heating and cooling over long periods of time is dependent on air temperature, demands are increasing for analyses of existing temperature records and for projections. Weather and climate also have profound influences on energy transmission, particularly in the case of electric power lines sensitive to lightning, icing, and salt spray.

Another area of interest concerns the problem of how much of the world's energy needs can be met from natural sources, principally the sun and wind. A considerable waste of energy occurs through poor design and building construction and general disregard of the meteorological environment. Substantial climatic alterations are produced by human activities due to urbanization, usually requiring modified local weather forecasts. A close watch needs to be kept on such situations, particularly where large power parks with massed generation facilities might produce larger-scale weather alterations (Suomi, 1975).

The commission is also looking at the problems associated with attempts to bring meteorologically marginal areas into agricultural production. In addition, its interest extends to the growing need for specialized weather forecasts for a wide range of consumers, from shipping and aviation to sport and recreation. In the past, forecasting efforts have tended to be concentrated on the general public and the issue of weather warnings, but many specialized interest groups require more specific details and varying types of information.

Landsberg (1978) has presented a detailed discussion of these themes for future meteorological work and has also identified other areas needing further investigation. Despite the great deal of work done to date on the effects of weather and climate on the human body, there are still many gaps in knowledge, which may be reduced by collaboration between the WMO and the World Health Organization.

Landsberg also sees a critical need for the use of existing long climatological records for analysis of socio-economic impacts of climatic fluctuations. This is closely linked to the need for continuing work on the incorporation of meteorological information into economic models. Although this has been done with some success for agricultural production and with less precision for energy consumption, there are still major requirements for accurate modelling of the roles of weather and climate in fields such as packaging, transportation and distribution of goods, and general labour conditions. In the face of climatic fluctuations existing records will have to be exploited more fully for their predictive value to permit continued economic and social well-being. It is also interesting to note that the US Central Intelligence Agency has clearly recognized and stated the need to understand the magnitude of international threats which occur as a function of climatic change (CIA, 1974).

The whole question of climate modelling is seen by many meteorologists to be especially urgent in the light of events such as drought and famine in the Sahel, reduced grain crops around the world, and the increasingly critical dependence of an adequate world food supply on good weather and climate conditions (Hammond, 1974). At the same time there seems to be a growing realization that climatic predictions, whether based on modelling or on extrapolation, are nowhere near as reliable as might be hoped for. Therefore, attempts to understand what controls climatic processes and to construct models capable of predicting climate are giving enhanced significance and recognition to meteorological research.

Many other areas of meteorological investigation, already of major interest, are likely to require increasing attention in the future, particularly the problems of weather modification, atmospheric pollution and agricultural meteorology. In addition, the contributions of meteorologists to society in general seem likely to grow steadily through the benefits derived from research into severe local storms, atmospheric turbulence, and aeronautical meteorology, for example (American Meteorological Society, 1976b).

In the USA in particular industrial meteorology plays an important complementary role to government weather services (Epstein, 1976). Industrial meteorology started in the late 1940s with the provision of weather modification services, but the services now offered by industrial meteorologists cover specialized weather forecasting, air quality studies, flood forecasting, and a wide range of requirements for weather information and expertise from business, industry, public utilities and public authorities (Bates, 1976; Elliott, 1976; Wallace, 1976). Booker (1976) saw increasing markets for meteorological services and products, with the greatest opportunities in providing forecasts and meteorological data. He also pointed to the potential for development of a commercial weather service on a national basis, and emphasized that industrial meteorologists need to be alert for the numerous opportunities that are likely to arise. The profession of industrial meteorology seems set to expand perhaps on a world-wide basis, with the provision of a range of services previously unavailable in many countries. In doing so it will greatly increase the opportunities for trained meteorologists, it will bring specialized meteorological services to a broader spectrum of the community, and will thus add further to the importance of the meteorologist's role in society.

Major elements in any such growth are public relations and education. Schneider (1974a) put the view that many politicians, economists and other specialists were unaware of all the possibilities of the applications of meteorology to economic development. Five years later this observation still retains considerable validity. It is important, therefore, to encourage decision makers and the public at large to gain a better understanding of the important role which meteorology can play in economic and social life. Better understanding should in turn lead to more constructive criticism and more positive efforts to provide feedback from consumers to producers. The results would be improvements in meteorological services and increasingly effective application of those services to the problems facing modern society.

Perhaps the problems can be resolved into one of a need for education on both sides, although it appears that the educators may have to be the meteorologists themselves, with the assistance of professional advertising agencies. Clearly, they have much to offer to promote the maintenance or improvement of our way of life, but it is equally clear that they still have to sell themselves and their products. At the same time, potential customers have to be educated to the possibilities presented by the meteorological services available.

SUGGESTIONS FOR FURTHER READING

ACKERMANN, W. C., CHANGNON, S. A. and DAVIS, R. J., 1974, 'The new weather modification law for Illinois.' *Bull. Amer. Met. Soc.*, 55, pp. 745–50.

ADDERLEY, E. E., 1968, 'Cloud seeding in western Victoria in 1966.' *Aust. Met. Mag.*, 16, pp. 56–63.

ANON., 1975, *World Weather Watch: the plan and implementation programme, 1976–1979.* WMO-No. 418 (World Meteorological Organization, Geneva).

BAIER, W., 1974, 'The challenge to agriculture meteorology.' *WMO Bull.*, 23, pp. 221–4.

BAUM, W. A., 1975, 'The roles of universities and weather services in the education of meteorological personnel.' *Bull. Amer. Met. Soc.*, 56, pp. 226–8.

BIGG, E. K., BROWNSCOMBE, J. L. and THOMPSON, W. J., 1969, 'Fog modification with long-chain alcohols.' *J. Appl. Met.*, 8, pp. 75–82.

BERGGREN, R., 1975, *Economic benefits of climatological services*. Technical Note 145, WMO-No. 424 (World Meteorological Organization, Geneva).

BRETHERTON, F. P., 1977, 'Training and distribution of Ph.D.s. in meteorology.' *Bull. Amer. Met. Soc.*, 58, pp. 230–2.

CHANDLER, T. J., 1974, 'The management of climatic resources.' In *Conservation in Practice*. Eds. Warren, A. and Goldsmith, F. B. (Wiley, London) pp. 15–35.

CROW, L. W., 1976, 'A trade association for industrial meteorologists.' *Bull. Amer. Met. Soc.*, 57, pp. 1333–4.

DAVIS, F. K., 1976, 'Weather and the media.' *Bull. Amer. Met. Soc.*, 57, pp. 1331–3.

DENNIS, A. S. and DRIEGE, D. F., 1966, 'Results of ten years of cloud seeding in Santa Clara County, California.' *J. Appl. Met.*, 5, pp. 684–91.

HAMMOND A. L., 1976, 'Concern over climate: researchers increasingly go public.' *Science*, 192, pp. 246–7.

HICKS, J. R., 1967, 'Improving visibility near airports during periods of fog.' *J. Appl. Met.*, 6, pp. 39–42.

HOLROYD, E. W., 1978, 'The practicability of dry ice for on-top seeding of convective clouds.' *J. Appl. Met.*, 17, pp. 49–63.

HORNIG, D. F., 1965, 'The atmosphere and the nation's future.' *Bull. Amer. Met. Soc.*, 46, pp. 438–42.

HOWELL, W. E., 1965, 'Twelve years of cloud seeding in the Andes of northern Peru.' *J. Appl. Met.*, 4, pp. 693–700.

JENSEN, C. E., 1975, 'A review of federal meteorological programs for fiscal years 1965–1975.' *Bull. Amer. Met. Soc.*, 56, pp. 208–24.

KESSLER, E., 1973, 'On the artificial increase of precipitation.' *Weather*, 28, pp. 188–94.

LANSFORD, H., 1977, 'Cloud seeding—no cure-all for weather problems.' *The Peanut Farmer*, 13, pp. 8–11.

MASON, B. J., 1970, 'Future developments in meteorology: an outlook to the year 2000.' *Quarterly J. Roy. Met. Soc.*, 96, pp. 349–68.

MAUNDER, W. J. and WHITMORE, A. D., 1969, 'The value of weather: challenge of assessment.' *Aust. Geogr.*, 11, pp. 22–8.

MILLER, A. A., 1956, 'The use and misuse of climatic resources.' *Adv. of Sci.*, 13, pp. 56–66.

O'MAHONEY, G., 1967, 'Cloud seeding in Wimmera-Mallee, Victoria, 1966.' *Aust. Met. Mag.*, 15, pp. 133–47.

OSMUN, W. G., 1969, 'Airline warm fog dispersal program.' *Weatherwise*, 22, pp. 48–53, 87.

OTT, W. R. and THOM, G. C., 1976, 'A critical review of air pollution index systems in the United States and Canada.' *J. Air Polln. Control Assoc.*, 26, pp. 460–70.

OTTAR, B., 1977, 'International agreement needed to reduce long-range transport of air pollutants in Europe.' *Ambio*, 6, pp. 262–9.

PETTERSSEN, S., 1966, 'Recent demographic trends and future meteorological services.' *Bull. Amer. Met. Soc.*, 47, pp. 950–63.

PLANK, V. G. and SPATOLA, A. A., 1966, 'Cloud modification by helicopter wakes.' *J. Appl. Met.*, 8, pp. 566–78.

SCHNAPF, A., 1977, 'A survey of the United States meteorological satellite programs.' *Weatherwise*, 30, pp. 180–91.

SONKA, S. T. and CHANGNON, S. A., 1977, 'A methodology to estimate the value of weather modification projects; an illustration for hail suppression.' *J. Appl. Met.*, 16, pp. 667–72.

STAGG, J. M., 1961, 'Meteorology and the community.' *Quarterly J. Roy. Met. Soc.*, 87, pp. 465–71.

STANDLER, R. B. and VONNEGUT, B., 1972, 'Estimated possible effects of silver iodide cloud seeding on human health.' *J. Appl. Met.*, 11, pp. 1388–91.

WALLEN, C. C., 1975, 'Global atmospheric monitoring.' *Environmental Sci. and Technology*, 9, pp. 30–4.

References

ABBEY, R. F., 1976, 'Risk probabilities associated with tornado windspeeds.' In *Proceedings of the Symposium on Tornadoes: Assessment of Knowledge and Implications for Man, June 22-24 1976, Lubbock, Texas.* Ed. Peterson, R. E. (Texas Tech University), pp. 177-236.

ACKERMANN, B., 1974a, 'Wind field over the St. Louis metropolitan area.' *J. Air Polln. Control Assn.*, 24, pp. 232-6.

— 1974b, 'METROMEX: wind fields over St. Louis in undisturbed weather.' *Bull. Amer. Met. Soc.*, 55, pp. 93-5.

ADAMS, R. L., 1973, 'Uncertainty in nature, cognitive dissonance and the perceptual distortion of environmental information weather forecasts and New England beach trip decisions.' *Econ. Geog.*, 49, pp. 287-97.

ALEXANDER, T., 1974, 'Ominous changes in the World's weather.' *Fortune*, Feb., pp. 90-5.

D'ALLENGER, P. K., 1970, 'Parachuting and the weather.' *Weather*, 25, pp. 188-91.

ALMQVIST, E., 1974, 'An analysis of global air pollution.' *Ambio*, 3, pp. 161-7.

AMERICAN METEOROLOGICAL SOCIETY, 1976a, 'Policy statement of the American Meteorological Society on weather forecasting.' *Bull. Amer. Met. Soc.*, 57, pp. 1460-1.

— 1976b, 'Atmospheric Sciences and the problems of society.' *Bull. Amer. Met. Soc.*, 57, pp. 199-212, 436-40 and 1447-54.

ANDERSON, L. G. and BURNHAM, J. M., 1973, 'Application of economic analyses to hurricane warnings to residential and retail activities in the U.S. Gulf of Mexico coastal region.' *Mon. Weath. Rev.*, 101, pp. 126-31.

ANDERSON, R. 1976, 'Weather and crop yields in north-east Scotland.' *Farm Management Rev.*, 8, pp. 33-45.

ANGELL, J. K. and KORSHOVER, J., 1977, 'Estimate of the global change in temperature, surface to 100 mb, between 1958 and 1975.' *Mon. Weath. Rev.*, 105, 375-85.

ANON., 1976a, 'Meteorology and ocean affairs.' *WMO Bull.*, 25, pp. 181-3.

— 1976b, 'The first GARP global experiment.' *WMO Bull.*, 25, pp. 185-8.

— 1977, 'Spray cans and the ozone layer.' *Ecos*, 14, pp. 3-9.

— 1978a, 'Rainmaking: the state of the art.' *Ecos*, 16, pp. 15-18.

— 1978b, 'World Climate Programme.' *WMO Bull.*, 27, pp. 203-5.

— 1978c, 'World Climate Programme.' *WMO Bull.*, 27, pp. 276-8.

— 1978d, 'Next year's weather—the largest scientific experiment ever.' *Ecos*, 18, pp. 3-7.

ANTHES, R. A., 1976, 'Numerical prediction of severe storms—certainty, possibility or dream?' *Bull. Amer. Met. Soc.*, 57 pp. 423–30.

ARNOLD, G. W. and BENNETT, D., 1975, 'The problem of finding the optimum solution.' In *Study of Agricultural Systems.* Ed. Dalton, G. E. (Applied Science Publishers Ltd., London) pp. 129–73.

ARNOLD, G. W. and CAMPBELL, N. A., 1972, 'A model of a ley farming system with particular reference to a sub-model for animal production.' *Proc. Aust. Soc. Animal Prodn.*, 9, pp. 23–30.

ARNOLD, G. W. and GALBRAITH, K. A., 1978, 'Climatic change and agriculture in western Australia.' In *Climatic Change and Variability: a Southern Perspective.* Eds. Pittock, A. B. et al. (CUP, Cambridge) pp. 297–300.

ATKINSON, B. W., 1968, 'A preliminary examination of the possible effect of London's urban area on the distribution of thunder rainfall 1951–60.' *Trans. Inst. Brit. Geogrs.*, 44, pp. 97–118.

— 1969, 'A further examination of the urban maximum of thunder rainfall in London.' *Trans. Inst. Brit. Geogrs.*, 45, pp. 97–119.

— 1970, 'Meso-systems in the atmosphere.' *Canad. Geogr.*, 14, pp. 286–308.

ATLAS, D., 1976, 'Overview: the prediction, detection and warning of severe storms.' *Bull. Amer. Met. Soc.*, 57, pp. 398–401.

AULICIEMS, A., 1977, 'Thermal comfort criteria for indoor design temperatures in the Australian winter.' *Architectural Science Rev.*, 20, pp. 86–9.

AULICIEMS, A. and DE FREITAS, C. R., 1976, 'Cold stress in Canada: a human bioclimatic classification.' *Internl. J. Biomet.*, 20, pp. 287–94.

AUSTRALIAN ACADEMY OF SCIENCE, 1976, *Report of a Committee on Climatic Change.* Report No. 21.

BACASTOW, R. B., 1976, 'Modulation of atmospheric carbon dioxide by the Southern Oscillation.' *Nature*, 261, pp. 116–18.

BACH, W., 1976a, 'Changes in the composition of the atmosphere and their impact upon climatic variability—an overview.' *Bonner Meteorologische Abhandlungen*, 24.

— 1976b, 'Global air pollution and climatic change.' *Revs. Geophys. and Space Phys.*, 14, pp. 429–74.

BALL, D. J., 1976a, 'Photochemical ozone in the atmosphere of Greater London.' *Nature*, 263, pp. 580–2.

— 1976b, 'An air pollutant emission inventory for the Greater London area.' *Clean Air*, 10, pp. 7–9.

BARRIE, L. A., WHELPDALE, D. M. and MUNN, R. E., 1976, 'Effects of anthropogenic emissions on climate: a review of selected topics.' *Ambio*, 5, pp. 209–12.

BARRY, R. G., 1970, 'A framework for climatological research with particular reference to scale concepts.' *Trans. Inst. Brit. Geogrs.*, 49, pp. 61–70.

BATES, C. B., 1976, 'Industrial meteorology and the American Meteorological Society—a historical overview.' *Bull. Amer. Met. Soc.*, 57, pp. 1320–7.

BATES, M., 1966, 'The role of weather in human behavior.' In *Human Dimensions of Weather Modification.* Ed. Sewell, W. R. D., Research Paper No. 105 (Uni. of Chicago, Dept. of Geography) pp. 393–407.

BEDARD, A. J., HOOKE, W. H. and BERAN, D. W., 1977, 'The Dulles Airport pressure jump detector array for gust front detection.' *Bull. Amer. Met. Soc.*, 58, pp. 920–6.

BEEBE, R. G., 1967, 'The construction industry as related to weather.' *Bull. Amer. Met. Soc.*, 48, p. 409.

— 1970, 'TV weathercaster ratings—professional vs non-professional.' *Bull. Amer. Met. Soc.*, 51, pp. 399–401.

BELDING, H. S. and HATCH, T. J., 1955, 'Index for evaluating heat stress in terms of resulting physiological strain.' *Heating, Piping and Air Conditioning*, 27, pp. 129–36.

BELL, G. J., 1976, 'Seasonal forecasts and northern hemisphere circulation anomalies.' *Weather*, 31, pp. 282–92.

BERAN, D. W., HOOKE, W. H., LITTLE, C. G., and COONS, F., 1977, 'Airport weather service: some future trends.' *Bull. Amer. Met. Soc.*, 58, pp. 1182–6.

BERGEN, W. R. and MURPHY, A. H., 1978, 'Potential economic and social value of short-range forecasts of Boulder windstorms.' *Bull. Amer. Met. Soc.*, 59, pp. 29–44.

BIGG, K., 1977, 'Air outback—clean but not virginal.' *Ecos*, 11, pp. 23–6.

BLAINEY, G., 1977, 'The Great Drought.' *The National Times*, Oct. 31–Nov. 5, pp. 30–4, Sydney.

BOECK, W. L., SHAW, D. T. and VONNEGUT, B., 1975, 'Possible consequences of global dispersion of krypton 85.' *Bull. Amer. Met. Soc.*, 56, p. 527.

BOLIN, B., 1975, 'A critical appraisal of models for the carbon cycle.' In *The Physical Basis of Climate and Climate Modelling*, GARP Publications Series, No. 16, ICSU–WMO.

BOOKER, D. R., 1976, 'The future of industrial meteorology.' *Bull. Amer. Met. Soc.*, 57, pp. 1337–40.

BORNSTEIN, R. D., 1968, 'Observation of the urban heat island effect in New York City.' *J. Appl. Met.*, 7, pp. 575–82.

BOSART, L. F., 1976, Weather forecasting and analysis. In 'Atmospheric sciences and problems of society'. *Bull. Amer. Met. Soc.*, 57, pp. 200–2.

BOULOUX, C. and RUFFIÉ, J. (Eds.), 1971, *Pré-adaptation et adaptation génétique*, (Institut national de la Santé et de la Recherche médicale, Paris).

BRAND, S. and BLELLOCH, J. W., 1975, 'Cost effectiveness of typhoon forecast improvements.' *Bull. Amer. Met. Soc.*, 56, pp. 352–61.

BRIDGER, C. A. and HELFAND, L. A., 1968, 'Mortality from heat during July 1966 in Illinois.' *Internl. J. Biomet.*, 12, pp. 51–70.

BROECKER, W. S., 1975, 'Climatic change: are we on the brink of a pronounced global warming?' *Science*, 189, pp. 460–3.

BROMLEY, E., 1976, Aeronautical meteorological problems. In 'Atmospheric sciences and problems of society'. *Bull. Amer. Met. Soc.*, 57, pp. 439–40.

— 1977, 'Aeronautical meteorology: progress and challenges—today and tomorrow.' *Bull. Amer. Met. Soc.*, 58, pp. 1156–60.

BROOME, M. R., 1966, 'Weather forecasting and the contractor.' *Weather*, 21, pp. 406–10.

BRYSON, R. A., 1973, 'Drought in Sahelia: who or what is to blame?' *Ecologist*, 3, pp. 366–71.

— 1974a, 'A perspective on climatic change.' *Science*, 184, pp. 753–60.

— 1974b, *World climate and world food systems, III. The lessons of climatic history*, Institute for Environmental Studies Report 27, (University of Wisconsin, Madison).

— 1975, 'The lessons of climatic history.' *Environmental Conservation*, 2, pp. 163–70.

— 1978, 'Cultural, economic and climatic records.' In *Climatic Change and Variability: a Southern Perspective*. Eds. Pittock, A. B. et al. (CUP), pp. 316–27.

BRYSON, R. A. and MURRAY, T. J., 1977, *Climates of Hunger*. (A.N.U. Press, Canberra).

BUDYKO, M. I. and KAROL, I. L., 1975, 'Man's impact on the global climate.' In *Proceedings WMO/IAMAP Symposium on Long-Term Climatic Fluctuations, Norwich, 18–23 August 1975*. WMO–No. 421, pp. 465–71.

BUGAEV, V. A., 1973, 'World Weather Watch: its impact on economic and social development.' In *WMO Technical Note* 130, WMO–No. 370 (World Meteorological Organization, Geneva), pp. 91–100.

BUICK, T. R., McMULLAN, J. R., MORGAN, R. and MURRAY, R. B., 1976, 'On monitoring wind power.' *Weather*, 31, pp. 412–16.

BUREAU OF METEOROLOGY, AUSTRALIA, 1977, *Report on Cyclone Tracy, December 1974* (Australian Government Publishing Service, Canberra).

BURROWS, C. J., 1975, 'Late Pleistocene and Holocene moraines of the Cameron Valley, Arrowsmith Range, Canterbury, New Zealand.' *Arctic and Alpine Res.*, 7, p. 125.

CATCHPOLE, A. J. W. and MILTON, D., 1976, 'Sunnier prairie cities—a benefit of natural gas?' *Weather*, 31, pp. 348–54.

CESS, R. D., 1976, 'Climate change: an appraisal of atmospheric feedback mechanisms employing zonal climatology.' *J. Atmos. Sci.*, 33, pp. 1831–43.

CHANDLER, T. J., 1965, *The Climate of London* (Hutchinson, London).

— 1976, *Urban climatology and its relevance to urban design*. Technical Note No. 149, WMO–No. 438 (World Meteorological Organization, Geneva).

CHANDLER, T. J. and MUSK, L. F., 1976, 'The atmosphere in perpetual motion.' *Geogrl. Mag.*, 49, pp. 93–102.

CHANG, J.-H., 1970, 'Potential photosynthesis and crop productivity.' *Annals Assoc. Amer. Geogrs.*, 60, pp. 92–101.

— 1972, *Atmospheric Circulation Systems and Climates* (Oriental Publishing Co., Honolulu).

CHANGNON, S. A., 1973, 'Weather modification in 1972: up or down?' *Bull. Amer. Met. Soc.*, 54, pp. 642–6.

— 1975, 'The paradox of planned weather modification.' *Bull. Amer. Met. Soc.*, 56, pp. 27–37.

CHARAK, M. T. and DiGIULIAN, M. T., 1974, 'A review of federal legislation on weather modification.' *Bull. Amer. Met. Soc.*, 55, pp. 755–8.

CIA, 1974, *A Study of Climatological Research as it Pertains to Intelligence Problems* (US Central Intelligence Agency, Office of Research and Development).

CLARKE, J. F., 1972, 'Some effects of the urban structure on heat mortality.' *Environmental Res.*, 5, pp. 93–104.

CLARKE, J. F. and BACH, W., 1971, 'Comparison of comfort conditions in different urban and suburban microenvironments.' *Internl. J. Biomet.*, 15, pp. 41–54.

CLEARY, G., 1974, 'WHO activities in air pollution control.' *Clean Air*, 8, pp. 26–9.

CORNISH, P. M., 1977, 'Changes in seasonal and annual rainfall in New South Wales.' *Search*, 8, pp. 38–40.

COURT, A., 1976, 'Tornado damage probabilities.' In *Proceedings of the Symposium on*

Tornadoes: Assessment of Knowledge and Implications for Man, June 22–24, 1976, Lubbock, Texas. Ed. Peterson, R. E. (Texas Tech University), pp. 545–50.

Cox, R. A., Eggleton, A. E. J., Derwent, R. G., Lovelock, J. E. and Pack, D. H., 1975, 'Long-range transport of photochemical ozone in north-western Europe.' *Nature*, 255, pp. 118–21.

Crawford, T. V., 1976, Atmospheric turbulence and diffusion. In 'Atmospheric sciences and problems of society'. *Bull. Amer. Met. Soc.*, 57, pp. 438–9.

Cressman, G. P., 1969, 'Killer storms.' *Bull. Amer. Met. Soc.*, 50, pp. 850–5.

Dale, R. F., 1976, Agricultural and forest meteorology. In 'Atmospheric sciences and problems of society'. *Bull. Amer. Met. Soc.*, 57, pp. 202–4.

Daniels, A. and Bach, W., 1976, Simulation of the environmental impact of an airport on the surrounding air quality. *J. Air Polln. Control Assn.*, 26, pp. 339–44.

Darkow, G. L., 1976, 'Tornado detection, tracking and warning.' In *Proceedings of the Symposium on Tornadoes: Assessment of Knowledge and Implications for Man, June 22–24, 1976, Lubbock, Texas*, Ed. Peterson, R. E. (Texas Tech University), pp. 243–7.

Davis, F. K., 1958, 'Ulcers and temperature changes.' *Bull. Amer. Met. Soc.*, 39, pp. 652–4.

Davis, R. J., 1974, 'Weather modification litigation and statutes.' In *Weather and Climate Modification*. Ed. Hess, W. N. (John Wiley and Sons, New York), pp. 767–86.

Derrick, E. H., 1965, 'The seasonal variation in asthma in Brisbane: its relation to temperature and humidity.' *Internl. J. Biomet.*, 9, pp. 239–53.

— 1966, 'The annual variation in asthma in Brisbane: its relation to the weather.' *Internl. J. Biomet.*, 10, pp. 91–9.

— 1969, 'The short-term variation in asthma in Brisbane: its relation to weather and other factors.' *Internl. J. Biomet.*, 13, pp. 295–308.

DiGiulian, M. T. and Charak, M.T., 1974, 'Survey of state statutes on weather modification.' *Bull. Amer. Met. Soc.*, 55, pp. 751–4.

Driscoll, D. M., 1971, 'The relationship between weather and mortality in ten major metropolitan areas in the United States, 1962–5.' *Internl. J. Biomet.*, 15, pp. 23–39.

Driscoll, D. M. and Landsberg, H. E., 1967, 'Synoptic aspects of mortality. A case study.' *Internl. J. Biomet.*, 11, pp. 323–8.

Duckham, A. N., 1974, 'Climate, weather and human food systems—a world view.' *Weather*, 29, pp. 242–51.

Duncan, C. N., 1977, 'Solar and wind power—some meteorological aspects.' *Weather*, 32, pp. 242–51.

Dyer, A. J., 1974, 'The effect of volcanic eruptions on global turbidity, and an attempt to detect long-term trends due to man.' *Quart. J. Roy. Met. Soc.*, 100, pp. 563–71.

— 1975, 'An international initiative in observing the global atmosphere.' *Search*, 6, pp. 29–33.

Dyer, T. G. J. and Tyson, P.D., 1977, 'Estimating above and below normal rainfall periods over South Africa, 1972–2000.' *J. Appl. Met.*, 16, pp. 145–7.

Eagleman, J. R., 1974, 'A comparison of urban climatic modifications in three cities.' *Atmos. Env.*, 8, pp. 1131–42.

Ebdon, R. A.,1975, 'The quasi-biennial oscillation and its association with tropospheric circulation patterns.' *Met. Mag.*, 104, pp. 282–97.

EDHOLM, O. G., 1966, 'Problems of acclimatisation in man.' *Weather*, 21, pp. 340–9.

ELLIOTT, R. D., 1976, 'History of industrial meteorology and weather modification.' *Bull. Amer. Met. Soc.*, 57, pp. 1318–20.

ELSOM, D. M., 1978, 'The changing nature of a meteorological hazard.' *J. of Met.*, 3, pp. 297–9.

EPSTEIN, E. S., 1976, 'NOAA policy on industrial meteorology.' *Bull. Amer. Met. Soc.*, 57, pp. 1334–7.

ESCAP, LRCS and WMO, 1977, *Guidelines for Disaster Prevention and Preparedness in Tropical Cyclone Areas* (Geneva/Bangkok).

FALLS, R., 1978, 'Warning systems and their effectiveness—tropical cyclone warning system, northern region.' Paper presented to ANU/NARU Seminar on Natural Hazards Management in Northern Australia, Darwin, 11–14 Sept., 1978.

FANKHAUSER, R. K., 1976, 'Ozone levels in the vicinity of 33 cities.' *J. Air Polln. Control Assn.*, 26, pp. 771–7.

FARHAR, B. C., 1975, 'Weather modification and public opinion in South Dakota, 1972 and 1973.' *J. Appl. Met.*, 14, pp. 702–9.

FARHAR, B. C. and MEWES, J., 1975, 'Weather modification decision making: state law and public response.' *J. Appl. Met.*, 14, pp. 694–701.

FAWCETT, E. B., 1977, 'Current capabilities in prediction at the National Weather Service's National Meteorological Center.' *Bull. Amer. Met. Soc.*, 58, pp. 143–9.

FRANCESCHINI, G. A., 1976, Atmosphere–Ocean interaction. In 'Atmospheric sciences and problems of society'. *Bull. Amer. Met. Soc.*, 57, pp. 1449–51.

FRANK, N. L. and HUSAIN, S. A., 1971, 'The deadliest tropical cyclone in history?' *Bull. Amer. Met. Soc.*, 52, pp. 438–44.

FRIDRIKSSON, S., 1973, 'Crop production in Iceland.' *Internl. J. Biomet.*, 17, pp. 359–62.

FUJITA, T. T., 1973, 'Tornadoes around the world.' *Weatherwise*, 26, pp. 56–62 and 78–83.

FUJITA, T. T. and BYERS, H. R., 1977, 'Spearhead echo and downburst in the crash of an airliner.' *Mon. Weath. Rev.*, 105, pp. 129–46.

FUJITA, T. T. and CARACENA, F., 1977, 'An analysis of three weather-related aircraft accidents.' *Bull. Amer. Met. Soc.*, 58, pp. 1164–81.

GAFFNEY, D. O., 1973, *Atlas of Australian Resources, Second Series, Climate* (Geographic Section, Australian Dept. of Minerals and Energy, Canberra).

— 1976, 'An analysis of meteorological parameters for tourism recreation and related outdoor activities in Australia.' Paper presented to 47th ANZAAS Congress, Hobart, May, 1976.

GAGGE, A. P., STOLWIJK, J. A. J. and HARDY, J. D., 1967, 'Comfort and thermal sensations and associated physiological responses at various ambient temperatures.' *Environment Res.*, 1, pp. 1–10.

GAGIN, A. and NEUMANN, J., 1974, 'Rain stimulation and cloud physics in Israel.' In *Weather and Climate Modification*. Ed. Hess, W. N. (John Wiley and Sons, New York), pp. 454–94.

GENTILLI, J., 1971a, 'Climatic fluctuations.' In *Climates of Australia and New Zealand Vol. 13, World Survey of Climatology*. Ed. Gentilli, J. (Elsevier, Amsterdam).

— 1971b, *Climates of Australia and New Zealand, Vol. 13, World Survey of Climatology.* (Elsevier, Amsterdam).

— 1976, 'Atmospheric factors in disasters.' Paper presented to Symposium on Natural Hazards in Australia, Canberra, May, 1976 (Australian Academy of Science).

GENTRY, R. C., 1974, 'Hurricane modification.' In *Weather and Climate Modification.* Ed. Hess, W. N. (John Wiley and Sons, New York), pp. 497–521.

GEORGE, R. E., NEVITT, J. S. and VERSSEN, J. A., 1972, 'Jet aircraft operations: impact on the air environment.' *J. Air Polln. Control Assn.*, 22, pp. 507–15.

GIBBS, W. J., 1972, *Mathematics in meteorology: theory and practice* (Bureau of Meteorology, Australian Government Publishing Service, Canberra).

GIBBS, W. J. and MAHER, J. V., 1967, *Rainfall deciles as drought indicators.* Bulletin No. 48, (Bureau of Meteorology, Melbourne).

GILLETTE, D. G., 1975, 'Sulfur dioxide and material damage.' *J. Air Polln. Control Assn.*, 25, pp. 1238–43.

GILPIN, A. 1978, *Air Pollution.* 2nd ed. (University of Queensland Press, Brisbane).

GLANTZ, M., 1977, 'The value of a long-range weather forecast for the West African Sahel.' *Bull. Amer. Met. Soc.*, 58, pp. 150–8.

GOLDEN, J. H., 1976, 'An assessment of windspeeds in tornadoes.' In *Proceedings of the Symposium on Tornadoes: Assessment of Knowledge and Implications for Man, June 22–24, 1976, Lubbock, Texas.* Ed. Peterson R. E., Texas Tech University, pp. 5–42.

GOLDEN, J. H. and ABBEY, R. F., 1976, Severe local storms. In 'Atmospheric sciences and problems of society'. *Bull. Amer. Met. Soc.*, 57,, pp. 436–8.

GOLDSMITH, J. R. and PERKINS, N. M., 1967, 'Seasonal variations in mortality.' In *Biometeorology, Vol. 2, Pt. 1.* Eds Tromp, S. W. and Weihe W. H. (Pergamon Press, Oxford), pp. 97–114.

GONZALEZ, R. R., NISHI, Y. and GAGGE, A. P., 1974, 'Experimental evaluation of standard effective temperature: a new biometeorological index of man's thermal discomfort.' *Internl. J. Biomet.*, 18, pp. 1–15.

GORDON, A. H., 1974, 'Accuracy of weather forecasts.' *Nature*, 252, pp. 294–5.

GRANT, L. O. and KAHAN, A. M., 1974, 'Weather modification for augmenting orographic precipitation.' In *Weather and Climate Modification.* Ed. Hess, W. N. (John Wiley and Sons, New York), pp. 282–317.

GRAY, W. M., 1975, 'Tropical cyclone genesis.' *Atmospheric Science Paper* 234 (Department of Atmospheric Science, Colorado State University, Fort Collins).

GREEN, J. S. A., 1977, 'The weather during July 1976: some dynamical considerations of the drought.' *Weather*, 32, pp. 120–6.

GREENBURG, L., FIELD, F., REED, J. I. and ERHARDT, C. L., 1964, 'Asthma and temperature change. An epidemiological study of emergency clinic visits for asthma in three large New York hospitals.' *Arch. Environmental Health*, 8, pp. 642–7.

— 1967, 'Asthma and temperature change.' In *Biometeorology, Vol. 2, Pt 1.* Eds Tromp, S. W., and Weihe, W. H., (Pergamon Press, Oxford), pp. 3–6.

GREGORCZUK, M. and CENA, K., 1967, 'Distribution of effective temperature over the surface of the Earth.' *Internl. J. Biomet.*, 11, pp. 145–50.

GRIBBIN, J., 1976a, *Forecasts, Famines and Freezes* (Wildwood House, London).

— 1976b, 'Climatic change and food production.' *Food Policy*, 1, pp. 301–12.

— (Ed.), 1978, *Climatic Change*, (CUP, Cambridge).

HAAS, J. E., 1968, 'Sociological aspects of human dimensions of the atmosphere.' In *Human Dimensions of the Atmosphere*. Eds Sewell, W. R. D. et al. (National Science Foundation, Washington DC), pp. 53–7.

— 1973, 'Social aspects of weather modification.' *Bull. Amer. Met. Soc.*, 54, pp. 647–57.

— 1974, 'Sociological aspects of weather modification.' In *Weather and Climate Modification*. Ed. Hess, W. N. (John Wiley and Sons, New York), pp. 787–811.

HAMMOND, A. L., 1974, 'Modeling the climate: a new sense of urgency.' *Science*, 185, pp. 1145–7.

HANNA, S. R. and GIFFORD, F. A., 1975, 'Meteorological effects of energy dissipation at large power parks.' *Bull. Amer. Met. Soc.*, 56, pp. 1069–76.

HANSEN, J. B., 1970, 'The relation between barometric pressure and the incidence of peripheral arterial embolism.' *Internl. J. Biomet.*, 14, pp. 391–7.

HANSEN, J. B. and PEDERSEN, S. A., 1972, 'The relation between barometric pressure and the incidence of perforated duodenal ulcer.' *Internl. J. Biomet.*, 16, pp. 85–91.

HARDESTY, R. M., MANDICS, P. A., BERAN, D. W. and STRAUCH, R. G., 1977, 'The Dulles Airport acoustic-microwave radar wind and wind shear measuring system.' *Bull. Amer. Met. Soc.*, 58, pp. 910–19.

HARNACK, R. P. and LANDSBERG, H. E., 1975, 'Selected cases of convective precipitation caused by the metropolitan area of Washington, DC.' *J. Appl. Met.*, 14, pp. 1050–60.

HEATHCOTE, R. L., 1976, 'Drought in Australia: some problems for future research.' Paper presented to Symposium on Natural Hazards in Australia, Canberra, May 1976. (Australian Academy of Science).

HESS, W. N., (Ed.) 1974, *Weather and Climate Modification* (John Wiley and Sons, New York).

HOLOPAINEN, E. O., 1977, 'Energy balance of the Earth.' *Aust. Met. Mag.*, 25, pp. 89–103.

HOLTON, J. R. and LINDZEN, R. S., 1972, 'An updated theory for the quasi-biennial cycle of the tropical stratosphere.' *J. Atmos. Sci.*, 29, pp. 1076–80.

HOUGHTON, D., 1969, 'Acapulco '68' *Weather*, 24, pp. 2–18.

HOUNAM, C. E., 1967, 'Meteorological factors affecting physical comfort (with special reference to Alice Springs, Australia).' *Internl. J. Biomet.*, 11, pp. 151–62.

HOUNAM, C. E., BURGOS, J. J., KALIK, M. S., PALMER, W. C. and RODDA, J. C. E., 1975, *Drought and Agriculture: report of WMO working group on assessment of drought.* WMO Technical Note TN–138, WMO–No. 392 (World Meteorological Organization, Geneva).

HOWE, G. M., 1962, 'Windchill, absolute humidity and the cold spell of Christmas 1961.' *Weather*, 17, pp. 349–58.

HUDA, A. K. S., GHILDYAL, B. P., TOMAR, V. S. and JAIN, R. C., 1975, 'Contribution of climatic variables in predicting rice yield.' *Agric. Met.*, 15, pp. 71–86.

HUGHES, G. H., 1967, 'Summers in Manchester.' *Weather*, 22, pp. 199–200.

HUNTINGTON, E., 1945, *Mainsprings of Civilization* (John Wiley and Sons, New York).

HUSCHKE, R. E. (Ed.), 1959, *Glossary of Meteorology* (American Met Soc., Boston).

JAUREGUI, E. and SOTO, C., 1967, 'Wet bulb temperature and discomfort index areal distribution in Mexico.' *Internl. J. Biomet.*, 11, pp. 21–8.

JENKINSON, A. F., 1975, 'The accuracy of the Meteorological Office long range weather forecasts.' *Weather*, 30, pp. 288–90.

JOHNSON, S. R. and MCQUIGG, J. D., 1974, 'Some useful approaches to the measurement of economic relationships which include climatic variables.' In *Climatic Resources and Economic Activity*. Ed. Taylor, J. A. (David and Charles), pp. 223–36.

JOHNSON, S. R., MCQUIGG, J. D., and ROTHROCK, T. P., 1969, Temperature modification and cost of electric power generation. *J. Appl. Met.*, 8, pp. 919–26.

JOINT ORGANIZING COMMITTEE FOR GARP, 1975, *The Physical Basis of Climate and Climate Modelling*, GARP Publications Series, No. 16, ICSU–WMO.

JOLLIFFE, I. T. and FOORD, J. F., 1975, 'Assessment of long-range forecasts.' *Weather*, 30, pp. 172–81.

JULIAN, P. R. and STEINBERG, R., 1975, 'Commercial aircraft as a source of automated meteorological data for GATE and DST.' *Bull. Amer. Met. Soc.*, 56, pp. 243–51.

KALMA, J. D., 1974, 'An advective boundary-layer model applied to Sydney, Australia.' *Boundary Layer Met.*, 6, pp. 351–61.

KALMA, J. D., MILLINGTON, R. J. and ASTON, A. R., 1974, 'An urban heat island.' *Ecos*, 1, pp. 19–21.

KEELING, C. D., 1977, 'Impact of industrial gases on climate.' In *Energy and Climate: Outer Limits to Growth?* (National Academy of Science, Washington, DC).

KEELING, C. D., BACASTOW, R. B., BAINBRIDGE, A. E., EKDAHL, C. A., GUENTHER, P. R. and WATERMAN, L. S., 1976, 'Atmospheric carbon dioxide variations at Mauna Loa Observatory, Hawaii.' *Tellus*, 28, pp. 538–51.

KEELING, C. D., ADAMS, J. A., EKDAHL, C. A. and GUENTHER, P. R., 1976, 'Atmospheric carbon dioxide variations at the South Pole.' *Tellus*, 28, pp. 552–64.

KELLOGG, W. W., 1977, 'Effects of human activities on global climate—part 1.' *WMO Bull.*, 26, pp. 229–40.

KELLOGG, W. W. and SCHNEIDER, S. H., 1974, 'Climate stabilization: for better or for worse?' *Science*, 186, pp. 1163–72.

KELLOGG, W. W., COAKLEY, J. A. and GRAMS, G. W., 1975, 'Effect of anthropogenic aerosols on the global climate.' In *Proceedings WMO/IAMAP Symposium on Long-Term Climatic Fluctuations, Norwich, August 1975*. WMO–No. 421, pp. 323–30.

KELLY, P. M. and WRIGHT, P. B., 1978, 'The European drought of 1975–76 and its climatic context.' *Progress in Phys. Geog.*, 2, pp. 237–63.

KIRK, T. H., 1974, 'The use of numerical forecasts.' *Met. Mag.*, 103, pp. 14–21.

KLEINER, B. C. and SPENGLER, J. D., 1976, 'Carbon monoxide exposures of Boston bicyclists.' *J. Air. Polln. Control Assn.*, 26, pp. 147–9.

KNIGHTING, E., 1976, 'The weather a week ahead.' *Nature*, 262, pp. 162–3.

— 1978, 'The European Centre for Medium Range Weather Forecasts.' *WMO Bull.*, 27, pp. 14–19.

KNOX, J. B. and MACCRACKEN, M. C., 1976, 'Concerning possible effects of air pollution on climate. *Bull. Amer. Met. Soc.*, 57, pp. 988–91.

KUETTNER, J. P. and PARKER, D. E., 1976, 'GATE: report on the field phase.' *Bull. Amer. Met. Soc.*, 57, pp. 11–26.

KUKLA, G. J. and KUKLA, H. J., 1974, 'Increased surface albedo in the northern hemisphere.' *Science*, 183, pp. 709–14.

KULKARNI, R. N., 1976, 'Ozone trend and stratospheric circulation over Australia.' *Quart. J. Roy. Met. Soc.*, 102, pp. 697–704.

KUTZBACH, J. E., 1976, 'The nature of climate and climatic variations.' *Quaternary Research*, 6, pp. 471–80.

LAMB, H. H., 1958, 'The occurrence of very high surface temperatures.' *Met. Mag.*, 87, pp. 39–43.

— 1974, 'The current trend of world climate—a report on the early 1970s and a perspective.' *Research Publication No. 3, CRU RP3* (Climatic Research Unit, University of East Anglia).

— 1975, 'Remarks on the current climatic trend and its perspective.' In *Proceedings WMO/IAMAP Symposium on Long-Term Climatic Fluctuations, Norwich, August 1975.* WMO–No. 421, pp. 473–7.

LAMBERT, G., 1968, *L'adaptation. Physiologie et psychologie de l'homme aux conditions de vie désertique*, (Hermann, Paris).

LANDSBERG, H. E., 1968, 'A comment on land utilization with reference to weather factors.' *Agric. Met.*, 5, pp. 135–7.

— 1971, 'Interaction of man and his atmospheric environment.' In *Meteorology as related to the Human Environment*, WMO Special Environmental Report No. 2, WMO No. 312, (World Meteorological Organization, Geneva), pp. 65–72.

— 1978, 'Some applications of meteorology and climatology.' *WMO Bull.*, 27, pp. 103–5.

LAUR, T. M., 1976, 'The world food problem and the role of climate.' *Eos*, 57, pp. 189–95.

LAURMANN, J. A., 1975, 'On the prospects for seasonal climate forecasting. *Bull. Amer. Met. Soc.*, 56, pp. 1084–8.

LAVE, L. B. and SESKIN, E. P., 1970, 'Air pollution and human health.' *Science*, 169, pp. 723–33.

— 1973a, 'An analysis of the association between U.S. mortality and air pollution.' *J. Amer. Statistical Assn.*, 68, pp. 284–90.

— 1973b, *Air pollution and human health.* (Resources for the Future, Washington, DC).

LEE, D. and RATCLIFFE, R. A. S., 1976, 'Objective methods of long-range forecasting using surface pressure anomalies.' *Weather*, 31, pp. 56–65.

LEICESTER, R. H. and REARDON, G. F., 1976, *Wind damage in Australia.* (CSIRO, Division of Building Research, Melbourne).

LEMON, L. R., DONALDSON, R. J., BURGESS, D. W. and BROWN, R. A., 1977, 'Doppler radar application to severe thunderstorm study and potential real-time warning.' *Bull. Amer. Met. Soc.*, 58, pp. 1187–93.

LE ROY, P. A., LAU, W. W. P. and HOLDER, G. D. S., 1976, 'The occurrence and control of photochemical smog in Melbourne.' In *Occurrence and Control of Photochemical Pollution, Proceedings of Symposium and Workshop Sessions, Smog '76, Macquarie University, February, 1976.* (Clean Air Society of Australia and New Zealand, Sydney).

LESLIE, L. M. and SMITH, R. K., 1978, 'The effect of vertical stability on tornado-genesis.' *J. Atmos. Sci.*, 35, pp. 1281–8.

LINACRE, E. T. and HOBBS, J. E., 1977, *The Australian Climatic Environment*, (John Wiley and Sons, Brisbane).

LINDEN, F., 1959a, 'Weather in business.' *The Conference Board Business Record*, 16, pp. 90–4.

— 1959b, 'Merchandising with the weather.' *The Conference Board Business Record*, 16, pp. 144–9.

— 1962, 'Merchandising weather.' *The Conference Board Business Record*, 19, pp. 15–16.

LIST, R., 1978, 'The WMO precipitation enhancement project—PEP.' *WMO Bull.*, 27, pp. 237–42.

LORENZ, E. N., 1969, 'The predictability of a flow which possesses many scales of motion.' *Tellus*, 21, pp. 289–307.

— 1970, 'The nature of the global circulation of the atmosphere: a present review.' In *The Global Circulation of the Atmosphere*. Ed. Corby, G. A., (Royal Meteorological Society, London), pp. 2–23.

— 1973, 'On the existence of extended range predictability.' *J. Appl. Met.*, 12, pp. 543–6.

LOWRY, W. P., 1977, 'Empirical estimation of urban effects on climate: a problem analysis.' *J. Appl. Met.*, 16, pp. 129–35.

LUDWIG, F. L. and DABBERDT, W. F., 1973, 'Effects of urbanization on turbulent diffusion and mixing depth.' *Internl. J. Biomet.*, 17, pp. 1–11.

McCORMICK, R. A., 1958, 'An estimate of the minimum possible surface temperature at the South Pole.' *Mon. Weath. Rev.*, 86, pp. 1–5.

McDERMOTT, W., 1961, 'Air pollution and public health.' *Scientific American.*, 205, pp. 49–57.

MacDONALD, G. J., 1975, 'Weather modification as a weapon.' *Technology Rev.*, 78, pp. 57–62.

McDONALD, J. R., 1976, 'Tornado-generated missiles and their effects.' In *Proceedings of the Symposium on Tornadoes: Assessment of Knowledge and Implications for Man, June 22–24, 1976, Lubbock, Texas.* Ed. Peterson, R. E. (Texas Tech. University), pp. 331–48.

MACFARLANE, A. and WALLER, R. E., 1976, 'Short term increases in mortality during heatwaves.' *Nature*, 264, pp. 434–6.

McGRATH, C. A., 1971, 'Observations of an urban heat island.' *Soc. N'castle Uni. Geog. Students J.*, 1, pp. 36–40.

MACHTA, L. and TELEGADES, K., 1974, 'Inadvertent large-scale weather modification.' In *Weather and Climate Modification*. Ed. Hess, W. N. (John Wiley, New York), pp. 687–725.

McINTYRE, A. J., 1973, 'Effect of drought on the economy.' In *The Environmental, Economic and Social Significance of Drought*. Ed. Lovett, J. V. (Angus and Robertson, Sydney), pp. 181–92.

McINTYRE, D. P., 1970, 'What's happening to the forecast industry?' *Bull. Amer. Met. Soc.*, 51, pp. 314–18.

McMAHON, T. A. and WEEKS, C. R., 1973, 'Climate and water use in Australian cities.' *Aust. Geogrl. Studies*, 11, pp. 99–108.

McQUIGG, J. D., 1970, 'Some attempts to estimate the economic response of weather information.' *WMO Bull.*, 19, pp. 72–8.

— 1974, 'The use of meteorological information in economic development.' In *Applications of Meteorology to Economic and Social Development* (Schneider, R., McQuigg,

J. D., Means, L. L. and Klyukin, N. K.), Technical Note No. 132, WMO–No. 375, (World Meteorological Organization, Geneva), pp. 7–61.

MANABE, S. and WETHERALD, R. T., 1975, 'The effects of doubling the CO_2 concentration on the climate of a general circulation model.' *J. Atmos. Sci.*, 32, pp. 3–15.

MARKHAM, S. F., 1947, *Climate and the Energy of Nations*. (OUP, New York).

MASON, B. J., 1940, 'Future developments in meteorology: an outlook to the year 2000.' *Quart. J. Roy. Met. Soc.*, 96, pp. 349–68.

— 1966, 'The role of meteorology in the national economy.' *Weather*, 21, pp. 382–93.

— 1970, 'Future developments in meteorology: an outlook to the year 2000.' *Quart. J. Roy. Met. Soc.*, 96, pp. 349–68.

— 1975, 'The GARP Atlantic Tropical Experiment.' *Nature*, 255, pp. 17–20.

MAUNDER, W. J., 1970, *The Value of the Weather*, (Methuen University Paperback, London).

— 1971, 'Atmospheric resources: their evaluation and value.' *Proceedings Sixth Geography Conference, N.Z. Geographical Society, Christchurch*, pp. 98–103.

— 1972, *National econoclimatic models: problems and applications*. Technical Note 208, (New Zealand Meteorological Service, Wellington).

— 1973, 'Weekly weather and economic activities on a national scale: an example using United States retail trade data.' *Weather*, 28, pp. 2–18.

— 1974, 'National econoclimatic models.' In *Climatic Resources and Economic Activity*. Ed. Taylor, J. A., (David and Charles), pp. 237–57.

— 1977a, 'Weather and operation decision-making: the challenge.' *Quarterly Predictions, New Zealand Institute of Economic Research*, September, 1977.

— 1977b, 'Weather and climate factors in forecasting national dairy productions.' In *Management of Dynamic Systems in New Zealand Agriculture*, D.S.I.R. Information Series No. 129, pp. 101–26.

— 1977c, 'Climatic constraints to agricultural production. Prediction and planning in the New Zealand setting.' *N.Z. Agricultural Sci.*, 11, pp. 110–19.

— 1977d, 'The economic, social and political impact of climatic variations on agricultural production.' Paper presented at the WMO/FAO Technical Conference on the Potential Benefits of Agricultural Meteorology, Rome, October, 1977.

— 1978a, Economic and political issues. In Ch. 7 'The effect of climatic change and variability on mankind', In *Climatic Change and Variability: a southern perspective*. Ed. Pittock, A. B. et al., (CUP), pp. 327–34

— 1978b, Forecasting pastoral production: the use and value of weather based forecasts and the implications to the transportation industry and the nation.' Paper presented to Symposium on Meteorology and Transport, New Zealand Meteorological Service, Wellington, October 1978.

— 1979, 'The present and potential use of weather information in marketing intelligence.' Paper presented at Symposium on Food Production and Climatic Change, 49th Congress of ANZAAS, Auckland, January 1979.

MAUNDER, W. J., JOHNSON, S. R. and McQUIGG, J. D., 1971a, 'The effect of weather on road construction: a simulation model.' *Mon. Weath. Rev.*, 99, pp. 939–45.

— 1971b, 'The effect of weather on road construction: applications of a simulation model.' *Mon. Weath. Rev.*, 99, pp. 946–53.

METEOROLOGICAL OFFICE, GB, 1972, *Meteorological Glossary*. (HMSO, London.)

MILES, M. K., 1977, 'Atmospheric circulation during the severe drought of 1975/76.' *Met. Mag.*, 106, pp. 154–64.

MILLER, W. H., 1968, 'Santa Ana winds and crime.' *Prof. Geogr.*, 20, pp. 23–7.

MINOR, J., 1976, In *Proceedings of the Symposium on Tornadoes: Assessment of Knowledge and Implications for Man, June 22–24, 1976, Lubbock, Texas.*, Ed. Peterson, R. E. (Texas Tech. University, Lubbock).

MINTZ, Y., 1954, 'The observed zonal circulation of the atmosphere.' *Bull. Amer. Met. Soc.*, 35, pp. 207–14.

MITCHELL, J. M., 1971, 'The effect of atmospheric aerosols on climate with special reference to temperature near the Earth's surface.' *J. Appl. Met.*, 10, pp. 703–14.

— 1972, 'The natural breakdown of the present interglacial and its possible intervention by human activities.' *Quaternary Research*, 2, pp. 436–45.

MIYAKODA, K., 1974, 'Numerical weather prediction.' *Amer. Scientist*, 62, pp. 564–74.

MOMIYAMA, M., 1968, 'Biometeorological study of the seasonal variation of mortality in Japan and other countries on the seasonal disease calendar.' *Internl. J. Biomet.*, 12, pp. 377–93.

MOORE, P. D., 1976, 'Higher ozone concentration over Britain.' *Nature*, 263, p. 546.

MOREL, P., 1973, 'Developments in techniques of observing the atmosphere.' In *Lectures Presented at the IMO/WMO Centenary Conferences*, Technical Note No. 130, WMO–No. 370, (World Meteorological Organization, Geneva), pp. 53–78.

MORTIMORE, K. O., 1976, 'The great drought of 1975–1976.' *J. of Met.*, 1, pp. 373–8.

MUENCH, H. S., 1976, 'Use of digital radar data in severe weather forecasting.' *Bull. Amer. Met. Soc.*, 57, pp. 298–303.

MURPHY, A. H., 1977a, 'On the misinterpretation of precipitation probability forecasts.' *Bull. Amer. Met. Soc.*, 58, pp. 1297–9.

— 1977b, 'The value of climatological, categorical and probabilistic forecasts in the cost–loss ratio situation.' *Mon. Weath. Rev.*, 105, pp. 803–16.

— 1978, 'Hedging and the mode of expression of weather forecasts.' *Bull. Amer. Met. Soc.*, 59, pp. 371–3.

MURPHY, A. H. and WINKLER, R. L., 1974, 'Probability forecasts: a survey of National Weather Service forecasters.' *Bull. Amer. Met. Soc.*, 55, pp. 1449–53.

NATIONAL ACADEMY OF SCIENCES, 1973, *Weather and Climate Modification: Problems and Progress*. (Committee on Atmospheric Sciences, National Research Council, National Academy of Sciences, Washington, DC).

— 1976, *Climate and Food: Climate Fluctuation and U.S. agricultural production*. (National Academy of Sciences, Washington, DC).

NATIONAL CAPITAL DEVELOPMENT COMMISSION, AUSTRALIA, 1977, 'Low energy house design for temperate climates.' *Technical Paper* No. 22.

NATIONAL CENTER FOR AIR POLLUTION CONTROL, 1967, *Seminar on human biometeorology*. Public Health Service Publication No. 999-AP-25, (US Public Health Service, Washington, DC).

NEIBURGER, M., 1973, 'Developments in weather modification.' In *Lectures Presented at the IMO/WMO Centenary Conferences*, Technical Note No. 130, WMO–No. 370, (World Meteorological Organization, Geneva), pp. 79–86.

NEWTON, C. W., MILLER, R. C., FOSSE, E. R., BOOKER, D. R. and McMANAMON, P., 1978, 'Severe thunderstorms: their nature and their effects on society.' *Interdisciplinary Science Reviews*, 3, pp. 71–85.

NEW ZEALAND METEOROLOGICAL SERVICE, 1976, *Provisional list of New Zealand and World Weather Extremes*. Technical Information Circular No. 152, (N.Z. Meteorological Service, Wellington).

NKEMDIRIM, L. C., 1975, 'The global atmospheric research programme and the geographer.' *Prof. Geogr.*, 27, pp. 227–30.

NOAA, 1973, *The influence of weather and climate on United States grain yields: bumper crops or droughts. A report to the Administrator.* (NOAA, Washington, DC).

OESCHGER, H., SIEGENTHALER, V., SCHOTTERER, V., and GUGELMAN, A., 1975, 'A box diffusion model to study the carbon dioxide exchange in nature.' *Tellus*, 27, pp. 168–92.

OGDEN, T. L., 1969, 'The effect on rainfall of a large steelworks.' *J. Appl. Met.*, 8, pp. 585–91.

OKE, T. R., 1973, 'City size and the urban heat island.' *Atmos. Env.*, 7, pp. 769–79.

— 1974, *Review of Urban Climatology, 1968–1973*. Technical Note No. 134, WMO–No. 383, (World Meteorological Organization, Geneva).

— 1976, 'The distinction between canopy and urban boundary layer heat islands.' *Atmosphere*, 14, pp. 268–71.

OKE, T. R. and EAST, C., 1971, 'The urban boundary layer in Montreal.' *Boundary Layer Met.*, 1, pp. 411–37.

OKE, T. R. and MAXWELL, G. F., 1975, 'Urban heat island dynamics in Montreal and Vancouver.' *Atmos. Env.*, 9, pp. 191–200.

OLSEN, A. R. and WOODLEY, W. L., 1975, 'On the effect of natural rainfall variability and measurement errors in the detection of seeding effects.' *J. Appl. Met.*, 14, pp. 929–38.

OLSON, W. P., 1969, 'Rat-flea indices, rainfall and plague outbreaks in Vietnam, with emphasis on the Pleiku area.' *Amer. J. Tropical Medical Hygiene*, 18, pp. 621–8.

— 1970, 'Rainfall and plague in Vietnam.' *Internl. J. Biomet.*, 14, pp. 357–60.

ORLANSKI, I., 1975, 'A rational subdivision of scales for atmospheric processes.' *Bull. Amer. Met. Soc.*, 56, pp. 527–30.

PADMANABHAMURTY, B., 1972, 'A study of biotropism of climate in two Canadian cities.' *Internl. J. Biomet.*, 16, pp. 107–17.

PALMÉN, E. and NEWTON, C. W., 1969, *Atmospheric Circulation Systems* (Academic Press, New York).

PARRY, H. D., 1977, 'Ozone depletion by chlorofluoromethanes? Yet another look.' *J. Appl. Met.*, 16, pp. 1137–48.

PATTERSON, R. M., 1975, 'Traffic flow and air quality.' *Traffic Engineering*, 45, pp. 14–17.

PAUL, A. H., 1972, 'Weather and the daily use of outdoor recreation areas in Canada.' In *Weather Forecasting for Agriculture and Industry*. Ed. Taylor, J. A. (David and Charles, Newton Abbot).

PAULHUS, J. L. H., 1965, 'Indian Ocean and Taiwan rainfalls set new records.' *Mon. Weath. Rev.*, 93, pp. 331–5.

PAULUS, H. J. and SMITH, T. J., 1967, 'Association of allergic bronchial asthma with certain air pollutants and weather parameters.' *Internl. J. Biomet.*, 11, pp. 119–27.

PEARMAN, G. I., 1977, 'The carbon dioxide climate problem: recent developments.' *Clean Air*, 11, pp. 21–6.

PEARMAN, G. I. and GARRATT, J. R., 1972, 'Global aspects of carbon dioxide.' *Search*, 3, pp. 67–73.

PEARSON, A., 1976a, 'Tornado prediction.' In *Proceedings of the Symposium on Tornadoes: Assessment of Knowledge and Implications for Man, June 22–24, 1976, Lubbock, Texas*. Ed. Peterson, R. E. (Texas Tech. University), pp. 237–41.

— 1976b, 'Operational forecasting and warning of severe local storms.' *Bull. Amer. Met. Soc.*, 57, pp. 420–3.

PERKINS, H. C., 1974, *Air Pollution*. (McGraw-Hill, New York).

PERRY, A. H., 1971a, 'Climatic influences on the development of the Scottish skiing industry.' *Scottish Geogrl. Mag.*, 87, pp. 197–201.

— 1971b, 'Econoclimate—a new direction for climatology.' *Area*, 3, pp. 178–9.

— 1972, Weather, climate and tourism. *Weather*, 21, pp. 199–203.

— 1976, 'The long drought of 1975–76.' *Weather*, 31, pp. 328–34.

PERRY, A. H. and WALKER, J. M., 1977, *The ocean–atmosphere system* (Longman, London).

PETERSON, J. T., 1969, *The Climate of Cities: A Survey of Recent Literature* (US Dept of Health, Education and Welfare, National Air Pollution Control Administration, Rayleigh, N.C.).

PETTY, M. T., 1963, 'Weather and consumer sales.' *Bull. Amer. Met. Soc.*, 44, pp. 68–71.

PIGRAM, J. J. J. and HOBBS, J. E., 1975, 'The weather, outdoor recreation and tourism.' *J. Physical Ed. and Recreation*, 46, pp. 44–5.

PITTOCK, A. B., 1972, 'How important are climatic changes?' *Weather*, 27, pp. 262–71.

— 1975, 'Climatic change and the patterns of variation in Australian rainfall.' *Search*, 6, pp. 498–504.

PITTOCK, A. B., FRAKES, L. A., JENSSEN, D., PETERSON, J. A. and ZILLMAN, J. W. (Eds), 1978, *Climatic Change and Variability: a Southern Perspective* (CUP, Cambridge).

PONTE, L., 1976, *The Cooling*. (Prentice-Hall, Englewood Cliffs).

POULTER, R. M., 1962, 'The next few summers in London.' *Weather*, 17, pp. 253–7.

RACKCLIFF, P. G., 1965, 'Summer and winter indices at Armagh.' *Weather*, 20, pp. 38–44.

RAMAGE, C. S., 1976, 'Prognosis for weather forecasting.' *Bull. Amer. Met. Soc.*, 57, pp. 4–10.

— 1978, 'Further outlook—hazy.' *Bull. Amer. Met. Soc.*, 59, pp. 18–21.

RASOOL, S. I. and SCHNEIDER, S. H., 1971, 'Atmospheric carbon dioxide and aerosols: effects of large increases on global climate.' *Science*, 173, pp. 138–41.

RATCLIFFE, R. A. S., 1974, 'The use of 500 mb anomalies in long range forecasting.' *Quart. J. Roy. Met. Soc.*, 100, pp. 234–44.

— 1977, 'A synoptic climatologist's view of the 1975/76 drought.' *Met. Mag.*, 106, pp. 145–54.

— 1978, 'The problem of short-term climatic forecasting.' In *Climatic Change and Variability: a southern perspective*, Ed. Pittock, A. B., et al. (CUP, Cambridge), pp. 339–49.

RECK, R. A., 1976, 'Stratospheric ozone effects on temperature.' *Science*, 192, pp. 557–9.

REED, R. J., 1977, 'The development and status of modern weather prediction (Bjerknes Memorial Lecture).' *Bull. Amer. Met. Soc.*, 58, pp. 390–400.

RIGBY, M., 1965, 'The evolution of international co-operation in meteorology (1654–1965).' *Bull. Amer. Met. Soc.*, 46, pp. 630–3.

RIORDAN, P., 1970, 'Weather extremes around the world.' *Earth Sciences Lab. Tech. Rep.*, 70–45–ES (US Army, Natick, Massachusetts).

ROACH, W. T., 1970, 'Weather and Concorde.' *Weather*, 25, pp. 254–64.

ROBERTSON, G. W., 1974, 'World Weather Watch and wheat.' *WMO Bull.*, 23, pp. 149–54.

ROBERTSON, N. G. and COULTER, J. D., 1973, *The calculation of degree-day totals above or below any base.* Technical Note 224 (New Zealand Meteorological Service, Wellington).

ROGELL, R. H., 1972, 'Weather terminology and the general public.' *Weatherwise*, 25, pp. 126–32.

ROSENWAIKE, I., 1966, 'Seasonal variation of deaths in the United States 1951–1960.' *J. Amer. Statistical Assn.*, 61, pp. 706–19.

ROY, M. G., HOUGH, M. N. and STARR, J. R., 1978, 'Some agricultural effects of the drought of 1975–76 in the United Kingdom.' *Weather*, 33, pp. 64–74.

ROYAL COLLEGE OF PHYSICIANS, 1970, *Air pollution and health* (Pitman Medical and Scientific Publishing, London).

RUSSELL, J. S., 1973, 'Yield trends of different crops in different areas and reflections on the sources of crop yield improvement in the Australian environment.' *J. Aust. Inst. Agric. Sci.*, 39, pp. 156–66.

RUSSO, J. A., 1966, 'The economic impact of weather on the construction industry of the U.S.A.' *Bull. Amer. Met. Soc.*, 47, pp. 967–72.

SALINGER, M. J. and GUNN, J. M., 1975, 'Recent climatic warming around New Zealand.' *Nature*, 256, pp. 396–98.

SAMUEL, G. A., 1972, 'Some meteorological and other aspects of hot air ballooning.' *Met. Mag.*, 101, pp. 25–9.

SARGENT, T. F. and TROMP, S. W. (Eds), 1964, *A survey of human biometeorology.* WMO Technical Note 65, WMO–No. 160 (World Meteorological Organization, Geneva).

DE SAUVAGE-NOLTING, W. J., 1968, 'Seasonal variations of birthrates of cancer patients.' *Internl. J. Biomet.*, 12, pp. 293–5.

SAWYER, J. S., 1975, 'Some aspects of air pollution on a global and continental scale.' *Clean Air*, 9, pp. 12–18.

SCHNEIDER, R., 1974, 'Introduction.' In *Applications of Meteorology to Economic and Social Development.* Technical Note No. 132, WMO–No. 375 (World Meteorological Organization, Geneva).

SCHNEIDER, S. H., 1974, 'The population explosion: can it shake the climate?' *Ambio*, 3, pp. 150–5.

— 1975, 'On the carbon dioxide-climate confusion.' *J. Atmos. Sci.*, 32, pp. 2060–6.

— 1977, 'Climate change and the world predicament.' *Climatic Change*, 1, pp. 21–43.

SCHNEIDER, S. H. and DENNETT, R. D., 1975, 'Climatic barriers to long-term energy growth.' *Ambio*, 4, pp. 65–74.

SCHWERDTFEGER, P. and LYONS, T. J., 1976, 'Wind Field studies in an urban environment.' *Urban Ecology*, 2, pp. 93–108.

SELLERS, W. D., 1965, *Physical Climatology*, (University of Chicago Press).

SEN, A. R., BISWAS, A. K. and SANYAL, D. K., 1966, 'The influence of climatic factors on the yield of tea in the Assam valley.' *J. Appl. Met.*, 5, pp. 789–800.

SEWELL, W. R. D., 1968, 'Emerging problems in the management of atmosphere resources: the role of social science research.' *Bull. Amer. Met. Soc.*, 49, pp. 326–36.

SEWELL, W. R. D., KATES, R. W. and PHILLIPS, L. E., 1968, 'Human response to weather and climate: geographical contributions.' *Geogrl. Rev.*, 58, pp. 262–80.

SHERWIN, R. J., 1974, Contribution to panel discussion on 'Long term goals for environmental control—what's desirable and possible in the year 2000?' *J. Air Polln. Control Assn.*, 24, pp. 747–9.

SHUMAN, F. G., 1978, 'Numerical weather prediction.' *Bull. Amer. Met. Soc.*, 59, pp. 5–18.

SILVERMAN, B. A. and WEINSTEIN, A. I., 1974, 'Fog.' In *Weather and Climate Modification.* Ed. Hess, W. N., (John Wiley and Sons, New York), pp. 355–83.

SIMPSON, J., 1976, 'The GATE aircraft program: a personal view.' *Bull. Amer. Met. Soc.*, 57, pp. 27–30.

SIMPSON, J. and DENNIS, A. S., 1974, 'Cumulus clouds and their modification.' In *Weather and Climate Modification.* Ed. Hess, W. N., (John Wiley and Sons, New York), pp. 229–81.

SMIC, 1971, *Inadvertent Climate Modification: Report of the Study of Man's Impact on Climate.* (M.I.T. Press, Cambridge, Massachusetts).

SMITH, E. J., 1974, 'Cloud seeding in Australia.' In *Weather and Climate Modification.* Ed. Hess, W. N., (John Wiley and Sons, New York), pp. 432–53.

SMITH, K., 1975, *Principles of Applied Climatology.* (McGraw-Hill, London).

SMITH, L. P., 1968, 'Forecasting annual milk yields.' *Agric. Met.*, 5, pp. 209–14.

SMITHSON, P. A. and BALDWIN, H., 1978, 'Wind chill in lowland Britain.' *Weather*, 33, pp. 463–74.

SOUTHERN, R. L., 1976, 'Utilization of tropical cyclone warning: can man respond to scientific progress?' Paper presented to Symposium on Natural Hazards in Australia, Canberra, May 1976. (Australian Academy of Science).

SOUTHERN, R. L. and MACNICOL, B. F., 1973, 'The key role of meteorology in urban planning.' Paper presented to Section 21: Geographical Sciences, 45th ANZAAS Congress, Perth, 1973.

SPILLANE, K. T., 1978, 'Atmospheric characteristics on high oxidant days in Melbourne.' *Clean Air*, 12, pp. 50–6.

SPODNIK, M. J., CUSHMAN, G. D., KERR, D. H., BLIDE, R. W. and SPICER, W. S., 1966, 'Effects of environment on respiratory function. Weekly studies on young male adults.' *Arch. Environmental Health*, 13, pp. 243–54.

STEADMAN, R. G., 1971, 'Indices of windchill of clothed persons.' *J. Appl. Met.*, 10, pp. 674–83.

STEELE, A. T., 1951, 'Weather's effect on the sales of a department store.' *J. Marketing*, 15, pp. 436–63.

STEPHENSON, P. M., 1963, 'An index of comfort for Singapore.' *Met. Mag.*, 92, pp. 338–45.

STERN, A. C. (Ed.), 1968, *Air Pollution*, (Academic Press, New York).

STEWART, H. N. M., SULLIVAN, E. J. and WILLIAMS, M. L., 1976, 'Ozone levels in central London.' *Nature*, 263, pp. 582–4.

STRETEN, N. A., 1977, 'Aspects of the year-to-year variation of seasonal and monthly mean station temperature over the southern hemisphere.' *Mon. Weath. Rev.*, 105, pp. 195–206.

SULAKVELIDZE, G. K., KIZIRIYA, B. I. and TSYKUNOV, V. V., 1974, 'Progress of hail suppression work in the U.S.S.R.' In *Weather and Climate Modification*. Ed. Hess, W. N., (John Wiley and Sons, New York), pp. 410–31.

SUOMI, V. E., 1975, 'Atmospheric research for the nation's energy program.' *Bull. Amer. Met. Soc.*, 56, pp. 1060–8.

TAYLOR, J., 1971, 'Curbing the cost of bad weather.' *New Scientist*, 50, pp. 560–3.

TERJUNG, W. H., 1968, 'Some thoughts on recreation geography in Alaska from a physio-climatic viewpoint.' *Calif. Geogr.*, 9, pp. 27–39.

TEWUNGWA, S., 1973, 'WMO assists developing countries.' In *Lectures Presented at the IMO/WMO Centenary Conferences*, Technical Note No. 130, WMO–No. 370, (World Meteorological Organization, Geneva), pp. 101–9.

THOMPSON, L. M., 1969, 'Weather and technology in the production of wheat in the United States.' *J. Soil and Water Conservation*, 24, pp. 219–24.

THORNES, J. E., 1977a, 'The effect of weather on sport.' *Weather*, 32, pp. 258–68.

— 1977b, 'Ozone comes to London.' *Progress in Phys. Geog.*, 1, pp. 506–17.

TOMLINSON, A. I. and NICOL, B., 1976, *Tornado reports in New Zealand 1961–1975*. Technical Note 229, (New Zealand Meteorological Service, Wellington).

TOUT, D. G., 1977, 'Effective temperature and the hot spell of June–July 1976. *Weather*, 32, pp. 67–72.

— 1978, Mortality in the June–July 1976 hot spell. *Weather*, 33, pp. 221–6.

TRENBERTH, K. E., 1975, 'A quasi-biennial standing wave in the southern hemisphere and interrelations with sea surface temperature.' *Quart. J. Roy. Met. Soc.*, 101, pp. 55–74.

TROMP, S. W., 1963, 'Human Biometeorology.' *Internl. J. Biomet.*, 7, pp. 145–58.

TROMP, S. W. and BOUMA, J., 1965, 'Effect of weather on asthmatic children in the eastern part of the Netherlands.' *Internl. J. Biomet.*, 9, pp. 233–8.

— 1972, 'Influence of meteorological factors on suicide and suicide attempts in the western part of the Netherlands.' *Internl. J. Biomet.*, 16, pp. 116–17.

TUCKER, G. B., 1975, 'Climate: is Australia's changing?' *Search*, 6, pp. 323–8.

— 1976, 'Research and services: differing attitudes within the science of meteorology.' *Weather*, 31, pp. 104–13.

TUCKER. G. W., 1965, 'The weather and the holiday-maker.' In *What is Weather Worth?* (Bureau of Meteorology, Melbourne).

TURNER, D. B., 1968, 'The diurnal and day-to-day variations of fuel usage for space heating in St. Louis, Missouri.' *Atmos. Env.*, 2, pp. 339–51.

UNESCO, 1963, *Environmental physiology and psychology in arid conditions. Reviews of research*. Arid Zone Research 22, (UNESCO, Paris).

VAN LOON, H. and WILLIAMS, J., 1976, 'The connection between trends of mean temperature and circulation at the surface, part I: winter.' *Mon. Weath. Rev.*, 104, pp. 365–80.

WALLACE, J. E., 1976, 'Starting an industrial weather service.' *Bull. Amer. Met. Soc.*, 57, pp. 1329–31.

WALLEN, C. C., 1974, *A brief survey of meteorology as related to the biosphere.* Special Environmental Report No. 4, WMO–No. 372, (World Meteorological Organization, Geneva).

WALLINGTON, C. E., 1968, 'Forecasting for gliding.' *Weather*, 23, pp. 236–45.

WATT, G. A., 1967, 'An index of comfort for Bahrein.' *Met. Mag.*, 96, pp. 321–7.

WATTS, A., 1967, 'The real wind and the yacht.' *Weather*, 22, pp. 23–32.

— 1968, 'Winds for the Olympics.' *Weather*, 23, pp. 9–22.

WEATHERLEY, M. L., 1974, 'Trends in smoke concentrations before and after the Clean Air Act of 1956.' *Atmos. Env.*, 8, pp. 295–6.

WICKHAM, P. G., 1966, 'Weather for gliding over Britain.' *Weather*, 21, pp. 154–61.

WILLETT, H. C., 1974, 'Recent statistical evidence in support of the predictive significance of solar-climatic cycles.' *Mon. Weath. Rev.*, 102, pp. 679–86.

WILLIAMS, G. D. V., 1971/72, 'Geographical variations in yield-weather relationships over a large wheat growing region.' *Agric. Met.*, 9, pp. 265–83.

WILSON, O., 1967, 'Objective evaluation of wind chill index by records of frostbite in the Antarctica.' *Internl. J. Biomet.*, 11, pp. 29–32.

WINKLESS, N. and BROWNING, I., 1975, *Climate and the Affairs of Men*, (Peter Davies Ltd., London).

WINSTANLEY, D., 1972, 'Sharav.' *Weather*, 27, pp. 146–60.

WOODWELL, G. M. and HOUGHTON, R. A., 1976, 'Biotic influences on the world carbon budget.' Unpublished presentation at the Dahlem Conference on Global Chemical Cycles and their Alteration by Man.

WORLD HEALTH ORGANIZATION, 1972, *Health Hazards of the Human Environment*, (WHO, Geneva).

WORLD METEOROLOGICAL ORGANIZATION, 1976, 'WMO statement on modification of the ozone layer due to man's activities and some possible geophysical consequences.' *WMO Bull.*, 25, pp. 59–63.

WRIGHT, G. R., 1975, 'Carbon monoxide in the urban atmosphere.' *Arch. Env. Health.*, 30, pp. 123–9.

WRIGHT, P. B., 1975, *An index of the Southern Oscillation.* Climatic Research Unit Report, CRU RP4, (University of East Anglia, Norwich).

— 1976, 'Assessment of long-range forecasts.' *Weather*, 31, pp. 91–2.

— 1977, 'Persistent weather patterns.' *Weather*, 32, pp. 280–5.

— 1978, 'The Southern Oscillation.' In *Climatic Change and Variability: a Southern Perspective.* Ed. Pittock, A. B., et al., (CUP, Cambridge), pp. 180–4.

WRIGHT, P. B. and FLOOD, C. R., 1973, 'A method of assessing long-range forecasts. *Weather*, 28, pp. 178–87.

ZILLMAN, J. W., 1977, 'The first GARP global experiment.' *Aust. Met. Mag.*, 25, pp. 175–213.

Index